Water Supply in Emergency Situations

NATO Science for Peace and Security Series

This Series presents the results of scientific meetings supported under the NATO Programme: Science for Peace and Security (SPS).

The NATO SPS Programme supports meetings in the following Key Priority areas: (1) Defence Against Terrorism; (2) Countering other Threats to Security and (3) NATO, Partner and Mediterranean Dialogue Country Priorities. The types of meeting supported are generally "Advanced Study Institutes" and "Advanced Research Workshops". The NATO SPS Series collects together the results of these meetings. The meetings are co-organized by scientists from NATO countries and scientists from NATO's "Partner" or "Mediterranean Dialogue" countries. The observations and recommendations made at the meetings, as well as the contents of the volumes in the Series, reflect those of participants and contributors only; they should not necessarily be regarded as reflecting NATO views or policy.

Advanced Study Institutes (ASI) are high-level tutorial courses intended to convey the latest developments in a subject to an advanced-level audience

Advanced Research Workshops (ARW) are expert meetings where an intense but informal exchange of views at the frontiers of a subject aims at identifying directions for future action

Following a transformation of the programme in 2006 the Series has been re-named and re-organised. Recent volumes on topics not related to security, which result from meetings supported under the programme earlier, may be found in the NATO Science Series.

The Series is published by IOS Press, Amsterdam, and Springer, Dordrecht, in conjunction with the NATO Public Diplomacy Division.

Sub-Series

A.	Chemistry and Biology	Springer
B.	Physics and Biophysics	Springer
C.	Environmental Security	Springer
D.	Information and Communication Security	IOS Press
E.	Human and Societal Dynamics	IOS Press

http://www.nato.int/science
http://www.springer.com
http://www.iospress.nl

Series C: Environmental Security

Water Supply in Emergency Situations

edited by

Yair Sharan

Interdisciplinary Center for Technological
Analysis and Forecasting,
Tel Aviv, Israel

Abraham Tal

Interdisciplinary Center for Technological
Analysis and Forecasting,
Tel Aviv, Israel

and

Harry Coccossis

University of Thessaly,
Volos, Greece

Springer

Published in cooperation with NATO Public Diplomacy Division

Proceedings of the NATO Advanced Research Workshop on
Supply of Water to Cities in Emergency Situations
Tel-Aviv, Israel
5–7 June 2007

A C.I.P. Catalogue record for this book is available from the Library of Congress.

ISBN 978-1-4020-6304-6 (PB)
ISBN 978-1-4020-6303-9 (HB)
ISBN 978-1-4020-6305-3 (e-book)

Published by Springer,
P.O. Box 17, 3300 AA Dordrecht, The Netherlands.

www.springer.com

Printed on acid-free paper

CONTENTS

Page

LIST OF PARTICIPANTS vii

ACKNOWLEDGMENTS xi

ABOUT THE EDITORS xiii

ABOUT THE INSTITUTES xv

INTRODUCTION xvii

CHAPTER I: THREATS TO WATER SUPPLY AND RISK MANAGEMENT

1.1 The Threat to Water Supply Systems: An Overview, *Yair Sharan* 3
1.2 On Water Risk Management and Security, *Amnon Einav* 9
1.3 Water Quality Security Management, *Jacobo Sack* 17
1.4 Organizational Aspects of Water Supply, *Harry Coccossis, Madeleine Theochari, K. Dimitriou, G. Kallis* 21

CHAPTER II: COUNTRIES' AND MAJOR CITIES' CASE STUDIES

2.1 Alternate Water Supplies in Emergencies, *Sarah Burn* 29
2.2 Consequences of a Big Flood on the Paris Water Supply System, *Bruno Nguyen* 35
2.3 An Overview of Water Supply and Distribution Systems in the Metropolitan City of Istanbul, *Orhan Yenigun, Burak Demirel* 43
2.4 The Water Resources in Romania: Perspectives for the Sustainable Development in Order to Supply Water to Cities in Emergency Situation, *Liviu–Daniel Galatchi* 49

CHAPTER III: SECURITY OF WATER SUPPLY

3.1 Security of Water Supply as a Catalyst for Conflict Resolution in the Middle East, *Asher Vaturi, Abraham Tal* 59
3.2 Chernobyl Experience for Water Protection in Case of Radiation-Related Terrorist Attacks, *Zlobenko Borys* 65
3.3 Drinking Water Quality Monitoring in Emergency Situations in the Republic of Armenia, *Vardan Tserunyan, Bardukh Gabrielyan* 71
3.4 Secure Water Supply as a Central Element of Sustainable Water Management: Systems Analysis Approach for the Gunung Sewu Region in Central Java, Indonesia, *Helmut Lehn, Anne Schildt, Jürgen Kopfmüller* 77
3.5 The Anthropogenic Pollution of Water Basins: The Issue of Water Supply System Safety, *Christina Hakopian, Diana Hakobyan* 87

CHAPTER IV: DETECTION, MONITORING, AND WARNING

4.1 Water Security: Overview on Technologies and Methods for Rapid Detection of Terror Events, *Jean Negreanu* 93
4.2 The Ecological Monitoring of Pollution and Modeling of Hydroecosystems, *Trahel Vardanian* 101

4.3 Real-Time Analysis for Early Warning Systems Based on Modeling Concept, *Petr Ingeduld, Evžen Zeman, Palle Lindgaard Jorgensen, Marsha Horsner* 105

4.4 Application of Remote Sensing (Optical and SAR) to Monitoring Water Resources, *Garo Mardirossian* 115

CHAPTER V: TREATMENT OF CONTAMINATED WATER

5.1 Experience of Using Energy-Effective Water Disinfection Devices, *Rashid Khaydarov, Renat Khaydarov, Bekhzod Yuldashev, Roger Olsen* 127

5.2 Use of Natural Materials for Municipal Wastewater Treatment *Gagik Torosyan, Aleksander Isakov, Aida Aleksanyan, Meri Zakaryan* 133

CHAPTER VI: LEGAL ASPECTS

6.1 Supply of Water to the Population of Azerbaijan in Emergency Situations and its Legal and Institutional Aspects, *Rafig Verdiyev* 141

6.2 Unification of Standards for Drinking Water from Different Sources: Preliminary Approach for Water Security in Israel, *Tzvi Levinson, Gil Dror* 159

SUBJECT INDEX 165

AUTHOR INDEX 173

LIST OF PARTICIPANTS

AIDA ALEKSANYAN
State Engineering University of Armenia
105 Teryan Street
375009, Yerevan
Republic of Armenia

SARAH BURN
London Resilience Team, Government Office for London
Riverwalk House, 157-161
Millbank, London, SW1P 4RR
England
E-mail: sburn.gol@go-regions.gsi.gov.uk
Tel: +44-207-2173744
Fax: +44-207-2173744

HARRY COCCOSSIS
University of Thessaly
44 Nikis Street, 151 23 Maroussi
Athens,
Greece
E-mail: enpl@env.aegean.gr
Tel: +30-210-6800051
Fax: +30-2106800053

BURAK DEMIREL
Institute of Environmental Sciences and Centre for Disaster Management
Bogazici University
34342 Bebek
Istanbul
Turkey

K. DIMITRIOU
University of Thessaly
44 Nikis Street, 151 23 Maroussi
Athens
Greece

GIL DROR
The Levinson Environmental Law Firm
6 Ben Gurion Ave.
Haifa, 35414
Israel
E-mail: gil@environment.co.il
Tel: +972-4-8525246
Fax: +972-4-8643827

AMNON EINAV
Interdisciplinary Center for Technological Analysis and Forecasting
Tel Aviv University, Raman Aviv
Tel Aviv 69978
Israel
E-mail: amnone@post.tau.ac.il
Tel: +972-3-6407561
Fax: +972-3-6410193

BARDUKH GABRIELYAN
Institute of Hydroecology and Ichthyology
Armenian National Academy of Sciences
49, Marshall Bagramian ave., apt. 3, Yerevan
Republic of Armenia
E-mail: gabrielb@sci.am
Tel: +374-1-525424
Fax: +374-1-523830

LIVIU-DANIEL GALATCHI
Ovidius University of Constanta
Department of Ecology and Environmental Protection
Faculty of Natural Sciences, str. Dezrobirii, nr. 114, bl. IS7, apt. 16, Mamaia Boulevard 124, RO 900241 Constanta – 4, jud. Constanta, Romania
E-mail: galatchi@univ-ovidius.ro
Tel: +40-724-216273

DIANA HAKOBYAN
Yerevan State Linguistic University
Yerevan, Armenia

CHRISTINA HAKOPIAN
Yerevan State University
Department of Physical Geography
1, Alek Manoukian Street
Yerevan, 375025
Republic of Armenia
E-mail: chakopian@ysu.am
Tel: + 374-1-554641

MARSHA HORSNER
DHI, Inc. 301 South State Street
Newtown, PA 18940, USA

PETR INGEDULD
DHI Hydroinform a.s.
Prague 10, Na Vrsich 5
Czech Republic

ALEKSANDER ISAKOV
State Engineering University of Armenia
105 Teryan Street
375009, Yerevan
Republic of Armenia

PALLE LINDGAARD JORGENSEN
DHI Water and Environment
Agern Alle 11, DK-2970,
Horsholm
Denmark

G. KALLIS
University of Thessaly
44 Nikis Street, 151 23 Maroussi, Athens,
Greece

RASHID KHAYDAROV
Institute of Nuclear Physics
Scientific Devices Designing Department
Ulugbek, Tashkent, 702132, Uzbekistan
E-mail: green@sarkor.uz

JÜRGEN KOPFMÜLLER
Institute for Technology Assessment and
Systems Analysis
Helmholtz Research Center Karlsruhe
Hermann-v-Helmholtz-Platz 1
D-76344 Eggenstein-Leopoldshafen
Germany
E-mail: kopfmueller@itas.fzk.de

HELMUT LEHN
Institute for Technology Assessment
and System Analysis
Helmholtz Research Center Karlsruhe
Forschungszentrum Karlsruhe GmbH
P.O. Box 3640
D-76021 Karlsruhe
Germany
E-mail: lehn@itas.fzk.de
Tel: +49-7247-823977
Fax: +49-7247-826045

TZVI LEVINSON
The Levinson Environmental Law Firm
6 Ben Gurion Ave.
Haifa, 35414
Israel
E-mail: tzvi@environment.co.il
Tel: +972-4-8525246
Fax: +972-4-8643827

GARO MARDIROSSIAN
Space Research Institute at the Bulgarian
Academy of Sciences
National Science Funds
6 Moskovska Street
P.O. Box 799, Sofia 1000
Bulgaria
E-mail: garo@orionbg.net
Tel: +359-2-9793351
Fax: +359-2-9813347

JACOB NEGRANU
Interdisciplinary Center for Technological
Analysis and Forecasting
Tel Aviv University, Raman Aviv
Tel Aviv 69978
Israel
E-mail: jacobneg@post.tau.ac.il
Tel: +972-3-6407578
Fax: +972-3-6410193

JEAN NEGREANU
The Interdisciplinary Center for Technological
Analysis and Forecasting
Tel Aviv University, Tel Aviv
Israel

BRUNO NGUYEN
EAU De Paris 9, rue Victor Schoelcher
75014 Paris, France
E-mail: nguyen@Eaudeparis.fr
Tel: +33-1-40489905
Fax: +33-1-43219115

ROGER OLSEN
Camp Dresser and McKee, Inc.
1331 17th Street, Suite 1200
Denver, CO 80202, USA

JACOBO SACK
9 Hagvurot St.
Kfar Saba 44209, Israel
E-mail: ysack@bezeqint.net
Tel: +972-9-7654714

ANNE SCHILDT
Department of Urban and Environmental
Sociology
Helmholtz-Center for Environmental Research
UFZ Leipzig
Permoserstraße 15, D-04318 Leipzig
E-mail: anne.schildt@ufz.de

YAIR SHARAN
Interdisciplinary Center for Technological
Analysis and Forecasting
Tel Aviv University, Raman Aviv
Tel Aviv 69978, Israel
E-mail: sharany@post.tau.ac.il
Tel: +972-3-6407573/4
Fax: +972-3-6410183

ABRAHAM TAL
Interdisciplinary Center for Technological Analysis and Forecasting
Tel Aviv University, Raman Aviv
Tel Aviv 69978, Israel
E-mail: atal@eng.tau.ac.il
Tel: +972-3-6407586
Fax: +972-3-6410193

MADELEINE THEOCHARI
University of Thessaly
44 Nikis Street, 151 23 Maroussi
Athens, Greece
E-mail: Madeleine_theochari@yahoo.co.uk
Tel: +30-210-6800051
Fax: +30-2106800053

GAGIK TOROSYAN
State Engineering University of Armenia
105 Teryan Street
375009, Yerevan
Republic of Armenia
E-mail: gtorosyan@seua.am

VARDAN TSERUNYAN
USAID's Sustainable Water Resources Management Project in Armenia
48 Tigran Mets Ave., apt. #6
Yerevan 375005
Republic of Armenia
E-mail: vtserunyan@yahoo.com
Tel: +374-1-550653

TRAHEL GERASIM VARDANIAN
Yerevan State University
Department of Physical Geography
1, Alek Manoukian Street
Yerevan, 375025
Republic of Armenia
E-mail: tvardanian_ysu@yahoo.com
Tel: +374-1-554641

ASHER VATURI
Interdisciplinary Center for Technological Analysis and Forecasting
Tel Aviv University, Raman Aviv
Tel Aviv 69978
Israel
E-mail: asherv@eng.tau.ac.il
Tel: +972-3-6407584
Fax: +972-3-6410193

RAFIG VERDIYEV
Scientific Research Institute on Hydrometeorology of Ministry of Ecology and Natural Resources
36, H. Aliyev Street, Apt. 52, 3rd Mikrodistrict
AZ1111 Baku
Azerbaijan
E-mail: rafig2000@mail.ru
Tel: +994 12-5131687
Fax: +994 12-4368550

ORHAN YENIGUN
Bogazici University
Institute of Environmental Sciences and Centre for Disaster Management
Bebek 34342, Istanbul
Turkey
E-mail: yeniguno@boun.edu.tr
Tel: +90-212-358-1540 – 2198
Fax: +90-212-257-5033

MERI ZAKARYAN
State Engineering University of Armenia
105 Teryan Street
375009, Yerevan
Republic of Armenia

EVZEN ZEMAN
DHI Hydroinform a.s
Na Vrsich 5, Prague 10
Czech Republic
E-mail: E.Zeman@dhi.cz
Tel: +420 267 227 111
Fax: +420 271 736 912

ZLOBENKO BORYS
Institute of Environmental Geochemistry
NAS & ME of Ukraine
Palladina Ave., 34A
Kyiv-142, UA 03142
Ukraine
E-mail: bpz@vent.kiev.ua
Tel: +380-44-424-03-29
Fax: +380-44-424-00-30

ACKNOWLEDGMENTS

This book represents the outcome of NATO-sponsored Advanced Research Workshop (ARW) on Supply of Water to Cities in Emergency Situation, held in Tel Aviv, June 5–June 7, 2005. The workshop was organized by the Interdisciplinary Center for Technological Analysis and Forecasting (ICTAF) at Tel Aviv University in cooperation with the University of Thessaly, Department of Planning and Regional Development.

The Israel Ministry of Science and the Municipality of Tel –Aviv, and especially the Municipal Water Company, cooperated with ICTAF in organizing the workshop. The workshop brought together water experts from 14 countries in Europe, the Middle East, and Asia.

We would like to extend our gratitude and thanks for the opportunity we had to carry out this workshop to the following:

First and foremost, the NATO Science for Peace and Security programme, which accepted our proposal and provided funding for the workshop

The Israel Ministry of Science for their cooperation

The Municipality of Tel Aviv and especially David Jackman, manager of the municipal water company

The Israel Water Commission and especially the Israel Water Commissioner Shimon Tal and the Adviser for Water Security Gen (Res.) Aharon Vardi

Madeleine Theochari of the University of Thessaly, Department of Planning and Regional Development

Springer Publishing Company, The Netherlands, for their valuable advice and assistance, and for publishing this book

All the experts from Israel and abroad who took the time and made the effort to come to Tel Aviv and contribute to the exchange of knowledge and ideas in this important aspect of the water sector

Last but not least, the administrative staff of ICTAF and especially Ricky Shamit and Talma Schechter for many hours of work before, during, and after the workshop, and in the technical editing of the materials in this book.

ABOUT THE EDITORS

YAIR SHARAN

Dr. Yair Sharan is the Director of the Interdisciplinary Center for Technological Analysis and Forecasting (ICTAF). He holds B.Sc. and M.Sc. degrees in Mathematics and Physics. He also holds a Ph.D. in Physics and graduated in the Weizmann Institute of Science. He served as Israeli Science Attaché in Germany between 1988 and 1992. He was a senior consultant to the Israeli Ministry of Science in 1992–1993, coordinated the relations of the University of Tel Aviv with the EU during 1993–2003. He was a senior researcher in ICTAF during 1992–2000 and is the Director of the Center since 2000. He is a co-director of the Israel Future Society.

His main fields of interest are research and science policy, technology foresight and forecasting, technology assessment, and security. Together with ICTAF researchers, he has participated in many EU projects like e-Living, OPET, Nano2Life, Platform Foresight, and others. He is a member of the Millenium Project, American Future Society, and EU Foresight Expert Group.

Publications in Israel include the "Israeli Foresight Study towards the 21st Century," "ICT Uptake in the Israeli Education System," "Occupation for all," "The Development of Free Time," and more.

He has special interests in the security field and has completed studies in topics like "Non Conventional Terrorism," "Psychological Deterrence," "Emerging Technologies," "Future Threats of Terrorism," "Issues in water security," and more. He is a member of the Scientific Oversight Committee of the Center for Biological Terrorism in Tel Aviv University.

ABRAHAM TAL

Abraham Tal holds a B.Sc. in Geography, Geology, and Climatology from the Hebrew University, Jerusalem and an M.Sc. in Geodetic Science from the Ohio State University, Columbus, Ohio. Following a long military career, he retired from the army in 1980 at the rank of Lt. Colonel and has been a senior researcher at the Interdisciplinary Center for Technological Analysis and Forecasting at Tel Aviv University since then. He is an expert in earth-related issues of space, remote sensing, and environment. He was, for many years, a consultant to the Israel Space Agency on remote sensing-related issues. He participated as the ICTAF representative in the EURTD 5th framework METRON project, studying sustainable use of water in Metropolitan Areas (e.g. "Tel-Aviv Case Study – Final Report" Tal, A., Elhannani, S. "Sustainable Use of Water in Metropolitan Areas" – METRON Project, Report to EU October 2000). He has carried out research projects for the Israel Water Commission, for the Ministry of Science and for the Ministry of Energy and Infrastructures, such as use of remote sensing for locating submarine water sources. He has carried out studies of the urban heat island (e.g. "Urban Heat Island Effects as Measured by Satellite and the Need for UHI Consideration in Urban Planning" Tal, A. European Commission Energy Management, Sustainable Energy Information Magazine No. 3 2002). He has represented ICTAF in the energy-related OPET network and in the EURTD LandWaterMED Network on the use of geomatics for the management of land and water in the Mediterranean partner countries. He has been involved in many projects related to advancing the Middle East peace process (e.g. "Development of a Spatially Referenced Database for Management of an Israeli–Palestinian Transboundary Watershed" Rishmawi, K.N., Albert, J., and Tal, A. Report to the US Embassy Office of the Science Councelor, August 2001).

HARRY COCCOSSIS

Harry Coccossis is Special Secretary on Strategic Planning at the Greek Ministry of Tourism and previously Secretary General of the Greek National Tourism Organization (since 2004). He is also Professor of Spatial and Environmental Planning with specialization in tourism planning, Department of Planning and Regional Development, University of Thessaly, Greece (since 2002). He was a Professor in the Department of Environmental Studies of the University of the Aegean (1986–2002) and a visiting professor at several European and other universities.

His academic degrees include Ph.D., City and Regional Planning, Cornell University, USA; Master of Urban Planning, California State Polytechnic University, Pomona, USA, and Diploma in Architecture-Engineering, National Technical University of Athens.

He has experience in extensive international scientific research. He has directed and participated in more than 50 national and international research programs of integrated multidisciplinary issues with respect to development, tourism, spatial planning, environmental management, and planning. He also has 20 years of experience as a scientific adviser in the Mediterranean Action Plan of the United Nations Environment Program (UNEP/MAP) for sustainable development, sustainable tourism, integrated environmental management, urban development and renaissance, etc.

He has been a consultant and scientific adviser in many international organizations (European Commission, OECD, World Bank, FAO, UNEP, UNESCO, etc.) for spatial planning and environmental management, coastal area management, planning for tourism and environment, insular development with assignments in Mediterranean and African countries (Albania, Croatia, Slovenia, Cyprus, Tunisia, Morocco, Lebanon, Israel, Syria, Malta, Kenya, Tanzania, Comores, etc.).

He has also been a consultant and scientific adviser for Greek Government (Ministry of Coordination/Secretariat of the National Spatial Planning and Environment Council, Ministry of National Economy, Ministry for Environment, Physical Planning and Public Works, Organization for the Planning and Environmental Protection of Athens, Greek National Tourism Organization) for international environmental issues, environmental planning, urban and spatial planning, tourism development, etc. He has represented Greece in international meetings and workshops (European Commission, United Nations, etc.).

He has written – solely or in collaboration – 12 books and more than 50 scientific articles or chapters in international journals and books on spatial planning and environment, tourism, sustainable development, insular development, etc. Furthermore, he has made a number of presentations and publications in international and Greek scientific conferences. He is a member of International and Greek scientific Associations and societies.

ABOUT THE INSTITUTES

INTERDISCIPLINARY CENTER FOR TECHNOLOGICAL ANALYSIS AND FORECASTING

Founded in 1971, Interdisciplinary Center for Technological Analysis and Forecasting (ICTAF) is a leading institute in technology forecasting, foresight, assessment, and long-term planning, and is very active in the international community.

The multidisciplinary center taps the expertise of world-class scientists at Tel Aviv University and other well-known research establishments to create a core body of researchers with unrivaled knowledge in a diverse range of fields in exact sciences and engineering, geography, economics, education and social sciences, information technology, and communications.

ICTAF functions as a think tank, working alongside its governmental or business clients to produce far-reaching conclusions that are drawn from a unique blend of academic research and market know-how.

ICTAF's mission includes:

- Helping policy makers reach informed decisions based on technology's role in the development of economy and society.
- Serving as a think tank for future policy planning in Israel and abroad.
- Harnessing the knowledge of Tel Aviv University's scientists and scholars for the benefit of the economy and society.
- Contributing to the efforts by Tel Aviv University to positively influence Israel's development.
- Enhancing its leadership in the field of multidisciplinary foresight – covering science, technology, economics, and society.

A scientific board, including senior members of the university faculty, closely monitors ICTAF's work.

THE DEPARTMENT OF PLANNING AND REGIONAL DEVELOPMENT OF THE UNIVERSITY OF THESSALY

The Department of Planning and Regional Development of the University of Thessaly (DPRD) is the first department in Greece offering an undergraduate professional degree in Planning and Regional Development. Graduates of the department are certified engineers qualified for employment under Greek law. The department offers a 5-year professional diploma, a 1-year master's degree, and a doctorate degree in Planning and Regional Development. This prospectus aims to brief students on issues related to their studies. It provides a short introduction to the Department and the field of Planning and Regional Development as well as a description of the University, its development, and information concerning the organization of the department, courses, and the academic activity record of the faculty.

Planning and Regional Development is a field addressing a vast range of subjects. Fundamental to this field is the organization of space, development policies, environmental issues, and social processes. As a result, the employment prospects for the Department graduates are promising.

DPRD programme of studies is designed with a clear orientation toward analysis, layout, and decision making needed to solve spatial problems from an engineering perspective. This emphasis demands a strong technical background. The course of study takes place over a 5-year period, during which students are educated on subjects including economics, development, environment, management, methods of analysis, and problem solving with the use of advanced tools and techniques. DPRD is in the unique position, as a new department, to create the necessary blending of social and natural scientific subjects with innovative methods of instruction.

In 1989, the first 30 students were enrolled in the Department. Today, 170 students have already graduated as engineers in Planning and Regional Development and have joined the Technical Chamber of Greece. In the academic year 2000–2001, 60 undergraduate students enrolled in the Department. In addition, 24 students have been accepted for the M.Sc. programme, and 9 students have been accepted for the doctoral programme. It is the Department's philosophy that students perform best, and are most successful, when there is a high ratio of teachers to students. With this in mind, DPRD keeps enrollment figures low.

DPRD retains a faculty of 37 members. This includes 20 full-time teaching and research faculty members and 17 part-time adjunct faculty members. The faculty consists of a diverse group from scientific disciplines that range from Urban and Regional Planning to Economics and Regional Science, Geography, Mathematics, Computer Science, Image Processing and Remote Sensing, Civil Engineering, and Social Sciences. Many have served at various levels of public administration and have recieved distinctions for their research, publications, and contributions to the planning profession.

INTRODUCTION

Water is one of the most essential elements for sustaining life. National, regional, and local authorities throughout the world are responsible to maintain necessary infrastructure and safeguard resources for an orderly uninterrupted supply of good quality, healthy, and safe, water for everyday needs of all the population. These needs, which are growing fast with economic growth, development, and rising prosperity include water for drinking as well as for sanitation, laundry, gardening, recreation, and other domestic uses. An adequate supply of water resources should be safeguarded also for all sectors of the economy and society including agriculture, industry, energy, tourism, ecosystem protection, and more.

Drinking water is again becoming a global issue from many perspectives. There are still parts of the globe which lack the necessary water resources for their basic needs, whether in terms of quantity or quality, or both. Demographic growth in several world regions is likely to increase pressures for the development of water resources and further exploitation of existing ones. Changes in production and consumption patterns are expected to aggravate further the pressures on the quantity and quality of water resources across the world. Rising standards of living, intensive agriculture, and new industrial processes lead not only to increasing competition for water use and rising costs of water provision, but also to mounting risks. It is widely acknowledged that there is still a lot to be achieved in the direction of properly managing drinking water resources in the context of a strategy toward sustainable development at all spatial levels, local, regional, national, and global.

Furthermore, in recent years, there have been growing concerns on the overall ability of modern societies to effectively safeguard water resources, in the sense of their readiness to face threats and risks from catastrophic events which could disrupt water supply. Such events can range from accidental to intentional pollution, from partial disruption of the water delivery system in a city to total destruction of the infrastructure necessary for water supply across a region or a country. Compared to other similar "public" resources such as electricity, the water sector lacks the perspective and tools (standards and plans) to respond to security incidents, as well as large natural disasters (e.g. earthquakes, floods, and tsunamis). A comprehensive, holistic approach to drinking water management is necessary.

Israel is a good example of such concerns about water security in a multiple perspective. It is located in a region which is dry, where water is scarce. Moreover, the water problem is accentuated by security threats, namely potential terrorist activities against the water supply system. This situation is expected to be even more problematic in the future, mainly because of demographic overuse of existing natural water sources. Twenty years from now, national water resources will not be sufficient to supply all necessary needs and is expected to fulfill only the domestic and municipal needs of the people, and water may not be available for agriculture and industry. Furthermore, even the urban and municipal needs might not be sustainable. This region thus experiences a wide spectrum of threats to security of water supply and can give rise to innovative solutions to these kinds of problems to be used worldwide.

Reliability of water supply in the future will depend on the development of new water resources which are termed "manufactured water." Desalination of seawater and saline water will be one of the major alternatives, and the reuse of treated sewage water will be another one. In Israel, "manufactured water" will amount to almost 50% of the total water resources in the next decade. This will make possible reliability of the water system and secure water supply to all potential customers in Israel. Sustainability, however, concerns also ensuring high water quality and safety of drinking water. Israel, as well as other countries, faces severe problems of contamination of water sources creating real dangers to citizens' health. Contamination can emerge either from man-made failures and natural disasters or from deliberate actions by terrorists. These circumstances call for a national preparedness and protection program which will enable it to respond to give early warning to such events in the best available way. The September 11, 2001, terrorist attacks in the USA have been one of the major catalysts to the awareness to these threats worldwide including Israel. In Israel, this was even more enhanced by a major failure in the main Israeli national water carrier that left large parts of the population without drinking water for more than 24 h, because of penetration of a high level of chlorine contamination into the system. These events made the potential threat clear and visible.

Events relating to physical destruction of any of the system parts can include disruption of operation of the water system, the power components, telecommunications, and control components, as well as damage to reservoirs, pipes and pumps, stations, water carriers, and others. Terrorist attacks using biological or chemical agents can result in contamination of the water system endangering public health and creating a heavy psychological impact. The controversy among experts on the efficiency of such attacks is mounting; nevertheless, the issue should be addressed.

In the Israeli case, these threats have made policy makers renew plans and programs related to water supply management during a crisis resulting from terrorist activities, war, natural disasters, and major failures in the system. The Israeli's "existential" need for water exposes the vulnerability of the supply system. Protecting it is a vital need.

The Israeli experience can be a good example of the issues involved and such knowledge can be shared with other countries who are, or may be, similarly exposed to such threats; reciprocally, the special problems developed and experienced by other countries and special solutions can be shared too, thus achieving better response programs worldwide. This was a major reason why the Workshop on Water Security, which was the basis for preparing the contributions included in this volume, was organized in Israel.

The objective of the NATO Workshop on Water Security has been to bring together experts from various countries and discuss issues concerning the threats on water supply and necessary measures and procedures needed to cope with these threats and enable secure and sustainable water supply.

The book consists of six chapters referring to different aspects of the problem.

Chapter 1 consist of four contributions which evaluate the threat to water resources and associated supply, treatment, and distribution infrastructures. Natural as well as man-made and deliberate threat to water supply is assessed. Main threats include environmental pollution and contamination, terrorism of all sorts, and various technological hazards. Necessary steps and management needs in crisis situations are presented.

Chapter 2 focuses on case studies which present evidences and best practices from different cities and countries. These include the London Resilience case of flood effect in the city of Paris and management practices of the water distribution system in the metropolitan city of Istanbul. A special paper is presented on the urban groundwater in Romania. These papers highlight the complexity of sustainable water supply and management in urban areas with large populations.

Chapter 3 addresses issues of water supply security at international, national, and regional levels. Subjects range from nuclear radiation contamination learning from the Chernobyl experience, water as catalyst to conflict resolution, and down to damages to water systems in the local level. A special contribution presents the case of a sustainable water management project in the region of central Java in Indonesia.

Chapter 4 presents technological solutions in the field of detection monitoring and warning – a significant need for every national, regional, and municipality protection program preparing for these kinds of threats. The papers include rapid detection of terror events, ecological monitoring, remote sensing, and real-time analysis.

Chapter 5 reviews techniques and devices for treatment of contaminated water.

Chapter 6 covers some legal aspects and standards related to drinking water supply and distribution.

In spite of over a century of societal concerns with drinking water issues, tremendous progress in technological solutions with water supply, delivery and treatment systems, and advances in regulatory frameworks including economic incentives and administrative systems which involve public and private sectors, drinking water is still a central issue in public policy worldwide at international, national, regional, and local levels. New challenges require new approaches which involve multiple perspectives on water supply, as a response to new emerging social concerns. This volume intends to contribute to the international debate on safety, security, and water from a perspective of sustainable development.

The editors

CHAPTER I

THREATS TO WATER SUPPLY AND RISK MANAGEMENT

1.1 THE THREAT TO WATER SUPPLY SYSTEMS: AN OVERVIEW

YAIR SHARAN
The Interdisciplinary Center for Technological Analysis and Forecasting At Tel Aviv University, Israel

Abstract: The threat on water sources and supply systems emerges from a wide spectrum of natural and human sources. The events of September 11 in the USA focused the awareness on deliberate attack on water systems, mainly the use of possible contaminants to contaminate water and put civilians at risk. The water authorities have to evaluate the threat and prepare a protection plan accordingly.

Keywords: water security, water safety, chemicals, biologicals, radiological contaminants, toxicity

1. Introduction

Emergency situations in water supply systems have been declared usually in extreme cases caused by nature or exceptional system failures. Storms, floods, and drought are some such events which risk water systems and call for relevant national preparedness to ensure safe water supply during times of disasters. Failures that resulted in contamination of water systems were also taken into account. However, in recent years, following the events of September 11, awareness has increased to the possibility of deliberate attacks on water systems with the intention to cause mass casualties and extensive psychological impact to the civilian population. Destruction of water supply systems can be achieved either by physically damaging critical components of the system (e.g. using explosives) or by introducing harmful poisonous contaminants into the systems which will effect innocent civilians, as well as animals or plants. These effects might cause temporary or permanent injury to people which might result, in certain cases, in death. Contamination can be caused by various kinds of materials including chemicals, biological agents, or radioactive materials, some of them easily available in the free market.

The use of chemicals, biological, or radiological contaminants gives certain advantages to terrorists and thus pose a real and heavy threat on the water supply system (see Sharan et al., 2000).

The main advantages to terrorists are:

(a) Secrecy – it is possible to produce and apply such contaminants, covertly delaying warning and alarm.
(b) Indefectibility – some of these materials are difficult to detect using present technologies; casualties will then give the first alarm.
(c) Small quantities – they can potentially cause large effects, especially when biologicals groups are used.
(d) Persistent contamination – some contaminants are very persistent and difficult to be decontaminated; effects can be long-lasting.
(e) Incubation time – in the case of biologicals, there is a time delay between the application and the onset of casualties; first casualties might appear days or weeks, later limiting the possibility of the protection system to react on time.

In this overview, we focus on the acute and rapid contamination of potable water resulting from hostile action. This issue is technically far more demanding than the usual slow contamination of the water table by chronic diffusion of sewage and industrial water. Other threat to the water system will be referred to in other papers in this book.

Y. Sharan et al. (eds.), Water Supply in Emergency Situations, 3–7.

2. Chemical Contaminants

The list of chemicals that can potentially be used to deliberately contaminate water and threat those who might use them is long. The literature is full with such lists (see, e.g. www.epa.gov/ssafewater/md.html) which demonstrate the complexity of the problem. Every country establishes its own reference list reflecting its unique security situation. The potential of an agent to be included in such a list depends on various parameters including:

(a) Toxicity of the agent.
(b) Solubility of the agent in water that defines the ability to disperse it in drinking water.
(c) Resistance of the material to hydrolysis or oxidation.
(d) Detectability of the agent in water (color, taste, smell, or other physical properties).
(e) Volatility of the agent – nonvolatile materials are more suitable to contaminate water sources.

Many materials suitable for terrorist needs can be easily purchased in the free market. Those (especially pesticides and insecticides) are commercial materials ready and accessible. Insecticides impact the nervous system similar to weaponized nerve agents, but are designed to be toxic for insects and less so for people unless injected or absorbed in massive doses. Parathion and the oxide carbonate aldicarb are two such insecticides commercially available. Parathion, in a very simple process, can be transferred into far more toxic paraoxon. Manufactured in relevant quantities this could be used to contaminate water reservoirs or introduced deliberately into water carriers. Accidents because of extensive use of herbicides for agriculture needs demonstrated the relevance of such scenarios.

Terrorists can also try and make use of materials from other sources, especially designer drugs and weaponized agents. Designer drugs are not included in the usual drug lists. They are potent and cheaper then "known" drugs (e.g. heroin) and economically attractive to produce (e.g. 3-methyltentanyl [3MF]). It is 1,000 times more potent than heroin and such potentiality can poison water tanks. It can be produced in relatively simple labs from easily available compounds. This rout demonstrates the possibility of a terrorist group to synthesize potent chemicals covertly, which is an important consideration for such a group.
Weaponized materials are of course a very worrisome threat.

In this context, one can take into account, for example, agents like cyanic compounds which are water-soluble and are known to be toxic. Some of them were used in the past as chemical warfare (see also BBC News, 2002). Table 1 lists several such compounds and their attributes.

TABLE 1. Properties of several cyanic compounds. (From Whelton et al., 2003.)

Chemical formula	Chemical name	Typical phase	Color descriptor	Odor descriptor	Water-soluble (yes/no)
HCN	Hydrogen cyanide	Gas or liquid	Colorless Yellow or brown	Bitter almond marzipan peach	Yes
NaCN	Sodium cyanide	Solid	White	Almond-like	Yes
KCN	Potassium cyanide	Solid	White	Almond-like	Yes
CNCl	Cyanogens	Gas or liquid	Colorless		Yes

The potential threat of these compounds to water systems was exposed in February 4, 2002, in Rome when four Moroccan civilians tried to poison the drinking water of the American Embassy by introduction of 5 kg of cyanic compounds (see Whelton et al., 2003).

More potent chemical agents known as chemical warfare like VX (nerve agent) or sulfur mustard (blister agent) are persistent and if available can be used to poison water sources. However, in contrast to the above these materials are more difficult to acquire or self-produce. Nevertheless, any protection plan should take them into account in its preparedness program.

Last group of materials worth mentioning are disinfection and cleansing materials like chlorine, chlorine dioxide, sodium hypochlorite, and more. Overdoses of these materials can also harm water sources and injure people.

3. Biological Contaminants

In nature, one can find great variety of biological contaminants including bacteria, fungi, viruses, rickettsial agents, and toxins. The first four are groups of pathogens which cause disease in humans as well as in animals and plants. Some pathogens are infectious and can pass from human to human.

Table 2 presents a list of biological contaminants which can be used to contaminate water systems – reservoirs and carriers (Burrows and Renner, 1999). Table 3 presents the threat from biotoxins. One can see that chlorination of water does not always give enough pro-

TABLE 2. Summary of threat potential of replicating agents. (From Burrows and Renner 1999.)

Agent/disease	Weaponized	Water threat	Stable in water	Chlorine tolerance[a]
Anthrax	Yes	Yes	2 years (spores)	Spores resistant
Brucellosis	Yes	Probable	20–72 days	Unknown
Cholera	Unknown	Yes	Survives well	Easily killed
Clostridium perfringens	Probable	Probable	Common in sewage	Resistant
Glanders	Probable	Unlikely	Up to 30 days	Unknown
Q fever	Yes	Possible	Unknown	Unknown
Salmonella	Unknown	Yes	8 days, fresh water	Inactivated
Shigellosis	Unknown	Yes	2–3 days	Inactivated, 0.05 ppm., 10 min
Tularemia	Yes	Yes	Up to 90 days	Inactivated, 1 ppm, 5 min
Typhus	Probable	Unlikely	Unknown	Unknown
Smallpox	Possible	Possible	Unknown	Unknown
Cryptosporidiosis	Unknown	Yes	Stable days or more	Resistant

[a]Ambient temperature, 1 ppm free available chlorine, 30 min or as indicated

TABLE 3. Summary of threat potential of biotoxins. (From Burrows and Renner, 1999.)

Biotoxin	Weaponized	Water threat	Stable in water	Chlorine tolerance[a]
Aflatoxin	Yes	Yes	Probably stable	Probably tolerant
Botulinum toxins	Yes	Yes	Stable	Inactivated, 6 ppm, 20 min
Microcystins	Possible	Yes	Probably stable	Resistant at 100 ppm
Ricin	Yes	Yes	Stable	Resistant at 10 ppm
Staphylococcal enterotoxins	Probable	Yes	Probably stable	Unknown
T-2 mycotoxin	Probable	Yes	Stable	Resistant

[a]Ambient temperature, 1 ppm free available chlorine, 30 min or as indicated

tection. However, biologicals are sensitive to hydrolysis which together with extensive chlorination will destroy great part of the agent and one would need large quantities of the agent to ensure effectivity. Contamination of a municipal water supply requires compensation for a significant dilution factor and hence quantities of biologicals which might be beyond what terrorists might find easy to acquire or transport. In countries in which water systems are well protected by detectors and other systems that will "discover" such hazardous results, these intentions will be even harder to accomplish.

To sum up, biologicals might be a real threat to those who like to use water systems vulnerabilities to achieve evil results. Some of these materials, on the one hand, are easy to acquire and use, but on the other hand, good preparedness can prevent such a threat from being realized.

4. Radioactive Contaminants

Small masses of radioactive material are highly toxic, lethal, and persistent. Some of them are soluble and can contaminate reservoirs of water. Even though difficult, terrorists possibly can acquire them from the nuclear industry fuel cycle. In particular, reactor fuel rods contain a highly toxic mix of many different elements, each with different radiochemical properties. Some are soluble and some readily bind to soil or budding surfaces.

Radioactive material emits several forms of ionizing radiation: positively charged alpha particles and negatively charged beta particles, as well as neutrons, gamma rays, and x-rays. In living tissue, ionization of cells breaks apart chromosomes, destroys some cells, produces toxins, and increases permeability of the cell membrane.

The different radionuclides have different chemical properties and concentrate in or attack different organs or parts of the body. Strontium, for example, is chemically similar to calcium and naturally concentrates in bones damaging marrow and blood cell production; cesium is similar to phosphorus and concentrates in body tissue.

Not all the radionuclides are long-lived. Some such as iodine-131 have very short half-lives and in a few weeks becomes harmless. Others, including the principal radioactive elements such as plutonium, uranium, cesium, strontium, and cerium are long-lived. Table 4 lists basic properties of several common radioactive elements present in reactor fuels (the exact composition depends on the type of reactor and how long the fuel which are soluble.

Radionuclides are also used as tracers in industry and medicine and in special power sources. As a consequence, small amounts of radioactive material can be obtained from pharmacies or chemical suppliers. It is doubtful that a credible amount of radioactive material can be accumulated from such sources to contaminate large sources. However, it can cause a significant psychological effect.

Dispersion of the radioactive material is nontrivial. For such a threat to be meaningful, a source large enough such that decontamination is not easily accomplished has to be considered. Material has thus to be put into the system either sprayed or poured as a liquid, or dispersed as

TABLE 4. Properties of radionuclides. (From Sharan et al., 2000.)

Element	Symbol	Half-life	Chem similarity	Notes
Iodine	^{131}I	8.0 days	Halogen	Soluble, concentrates in thyroid
Strontium	^{90}Sr	28.6 years	Calcium	Soluble, binds to soil, uptake in veg
Cesium	^{137}Cs	30.1 years	Potassium	Soluble, binds to soil, concentrated in tissue
Plutonium	^{239}Pu	24,100 years	Rare earth	Chemical availability low, physical trans
Zirconium	^{95}Zr	65 days	Refractory	Binds to soil and colloids

a soluble powder. Significant quantities need to be acquired and transported requiring special safety and protection arrangement from those who plan to use them.

5. Conclusion

The threat on the water system resulting from deliberate contamination using various toxic materials is concrete and real. Chemicals, biologicals, and radiological materials are available in the free market and terrorists can acquire them and try to disrupt water reservoirs and carriers.

Such an event can result in a great number of casualties and engender a significant psychological panic. Nevertheless, such attempts are hazardous and risky to the terrorists themselves; moreover, they need to overcome technical and operational gaps to succeed. It is quite possible to protect critical components of the system and make use of available technologies to monitor water quality, detect contamination and provide a warning which will prevent the ultimate realization of the threat. Some papers in this book are devoted to the necessary relations to achieve better security of our water supply systems.

References

1. Whelton, A.J., Jensen, J.L., Richards, T.E., Valdivia, R.M. (2003) The Cyanic Threat to Potable Water, US Army Center for Health Promotion and Preventive Medicine (USACHPPM).
2. BBC News (2002) “Cyanide plotters” face terror charges, February 21, 2002. (http://news.bbc.co.uk/2/hi/europe/1833646.stm)
3. Sharan, Y., Small, R., et al. (2000) Non Conventional Terrorism, The Interdisciplinary Center for Technological Analysis and Forecasting, E-107, June 2000. (In cooperation with Fondation pour la Recherche Stratégique, France, and The international Policy Institute for Counter Terrorism, Israel.)
4. Burrows, W.D. and Renner, S.E. (1999) Biological Warfare Agents as Threats to Potable Water, US Army Center for Health Promotion and Prevention Medicine. Aberdeen Proving Ground, Maryland.

1.2 ON WATER RISK MANAGEMENT AND SECURITY

AMNON EINAV
The Interdisciplinary Center for Technological Analysis and Forecasting (ICTAF), Tel Aviv University, Israel

Abstract: Water is important for the life and economy of the Mediterranean people. It is a scarce commodity. This situation worsens due to population increase, increase in standards of living, irrigation use expansion, and other causes created by changes in lifestyles. The principles of meeting the challenge of water needs on international, as well as national, scales are reviewed. The threats to water resources and the associated supply, treatment, and distribution infrastructures are evaluated. Necessary steps, to be taken by the water infrastructure operator, to counter the threat situation are given. Highlights of necessary development and some research work are also provided.

Keywords: water, security, emergency, assessments

1. Introduction

Water plays an important role in the life and economy of Mediterranean people. In this situation of scarcity, Mediterranean basin demands for water are high. Today, the region uses around 300 billion cubic meters per year.

Two thirds of Mediterranean countries now use over 500 m^3 per year per inhabitant mainly because of heavy use of irrigation. But these *per capita* demands are irregular and vary across a wide range – from a little over 100 to more than 1,000 m^3 per year. Globally, demand has doubled since the beginning of the 20th century and increased by 60% over the last 25 years.

These changes vary for different countries as seen in Fig. 1.

- Demand is growing slowly and tending to stabilize, where it is not actually starting to drop, in northern countries (in line with their demographic development) and in some southern countries (where requirements are limited by the shortage of water), e.g. Cyprus, Israel.
- Demand is growing in the other countries, but actually falling on a *per capita* basis. However, in some countries the demand per inhabitant is still growing – either because demand is still low (Algeria) or because water development schemes and use for irrigation particularly are developing more rapidly than the population (Lebanon, Libya, and Turkey).
- Demand for water supply, including the water losses in the transport and distribution system.

The challenge of meeting water needs on an international scale was well summarized in the following eleven principles as defined by the World Water Assessment Programme and provided as a contribution to the International Conference on Freshwater (Bonn, December 2001).

1. *Meeting basic needs*: to recognize that access to safe and sufficient water and sanitation are basic human needs and are essential to health and well-being, and to empower people, especially women, through a participatory process of water management.

Y. Sharan et al. (eds.), Water Supply in Emergency Situations, 9–15.

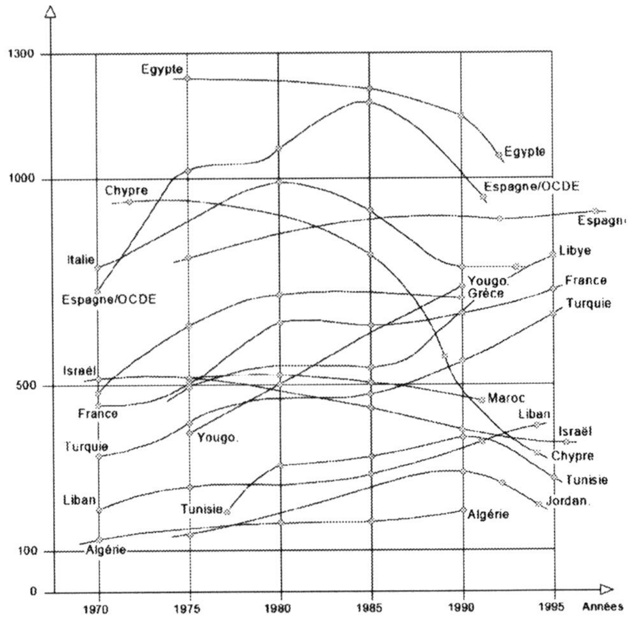

Figure 1. Evolutions of water demands per inhabitant, for all uses during past decades in Mediterranean countries in cubic meter per year. (From Margat and Vallee, 1999.)

2. *Securing the food supply*: to enhance food security, particularly of the poor and vulnerable, through the more efficient mobilization and use of water and the more equitable allocation of water for food production.
3. *Protecting ecosystems*: to ensure the integrity of ecosystems through sustainable water resources management.
4. *Sharing water resources*: to promote peaceful cooperation and develop synergies between different uses of water at all levels, whenever possible, within and – in the case of boundary and transboundary water resources – between states concerned, through sustainable river basin management or other appropriate approaches.
5. *Managing risks*: to provide security from floods, droughts, pollution, and other water-related hazards.
6. *Valuing water*: to manage water in a way that reflects its economic, social, environmental, and cultural values in all its uses, and to move towards pricing water services to reflect the cost of their provision. This approach should take account of the need for equity and the basic needs of the poor and the vulnerable.
7. *Governing water wisely*: to ensure good governance, so that the involvement of the public and the interests of all stakeholders are included in the management of water resources.
8. *Water and industry*: focuses on industry needs and the responsibility to respect water quality and take account of the needs of competing sectors.
9. *Energy and water*: recognizes that water is vital for all forms of energy production, and that there is a need to ensure that energy requirements are met in a sustainable manner.
10. *Ensuring the knowledge base*: reflects that good water policies and management depend upon the quality of knowledge available to decision makers.
11. *Water and cities*: acknowledges that urban areas are increasingly the focus of human settlements and economic activities, and that they present distinctive challenges to water managers.

The United Nations Millennium Declaration called upon all member states:

"To stop the unsustainable exploitation of water resources by developing water management strategies at the regional, national and local levels which promote both equitable access and adequate supplies."

Water is needed for all aspects of life. The general objective is to make certain that adequate supplies of water of good quality are maintained for the entire population, while preserving the hydrological, biological, and chemical functions of ecosystems, adapting human activities within the capacity limits of nature and eliminating carriers of water-related diseases. Scarcity of water is now recognized as a critical component of regional stability in many areas around the world.

Water resources and the associated *supply, treatment, and distribution infrastructures* are important elements of national, regional, and local security (Bib3) that may face a spectrum of threats. Recent threats suggest a trend that could put national, regional, or local water infrastructures at risk. Threats may come from Foreign hostile governments, terrorist groups, dissatisfied employees, local and transnational crime syndicates, vandals, natural disasters, and even unintentional man-made events such as adding of excess chemicals to the water system, all may pose potential threats to users of water systems. To counter the current threat situation, the water infrastructure operator must prepare:

- A systematic analysis of existing and emerging threats.
- Security enhancements based on the vulnerability assessments completed by water utilities.
- Early warning and response systems to detect and contain contaminants.
- An intervention action plan to mitigate the results of expected occurrences due to the threat.
- A recovery plan to bring water facilities back to their full operational capacity.

Water systems are critical to every community. Protection of public drinking water systems should be a high priority issue for state and for local authorities, for water system owners and operators, and also for water system designers and planners. Water security is not an end point, but a goal that can be achieved only through continued efforts to assess vulnerability of the water system and upgrade the water system emergency preparedness with the aim to ensure an uninterrupted water supply, which is essential for the protection of public health (safe drinking water and sanitation) (Bib6) and for safety (fire fighting).

Appropriate redundancy in the design and adequate security measures may help prevent loss of necessary services through water emergency situations such as drought, terrorist acts, vandalism, or induced, intentional or unintentional, contamination with chemicals or biological pollutants.

2. Some Security Provisions Related to Drought

Special provisions could be applied in case of drought, e.g. public entities, which supply water at retail, may be aloud to enforce a water conservation program with the aim to reduce the quantity of water used by the people therein, for the purpose of conserving the water supplies of such public entity.

A comprehensive *Emergency Water Management Plan* will have to be established to implement during this emergency case. This plan will be based upon the need to conserve water supplies during the emergency.

The water system manager or other assigned entity can be authorized and directed to implement the provisions of the agreed *Emergency Water Management Plan*, to ensure that water shall not be wasted or used unreasonably.

Under the water emergency measures: "No person shall knowingly use water or permit the use of water supplied by the water supplier for commercial, industrial, agricultural, governmental, or any other purpose, in a manner contrary to any provision as agreed between the water supply company and the water customer. No person shall use water in an amount in excess of the amounts authorized by the agreement, or during any period of time other than the periods of time specified in the agreement. At no time shall water be wasted or used unreasonably."

It is anticipated that the measures taken will result in a reduction in water use from a base period to be determined at the time of declaration of the water emergency situation. During the drought emergency, the following measures, or similar, shall apply except when reclaimed water or private well water is used following special permit.

3. Recommended Measures, to Cope with Water Shortages due to Drought

The *Water Emergency Board* will be activated and measures as described below be imposed:

1. All outdoor landscape irrigation will be prohibited.
2. Washing of autos, trucks, trailers, boats, airplanes, and other types of mobile equipment will be prohibited. Such washings will be exempted if the health, safety, and welfare of the public are contingent upon frequent vehicle cleaning such as garbage trucks and vehicles used to transport food and perishables (Gleick and IWRA, 1996).
3. Use of water for agricultural or commercial nursery purposes shall be permitted only under conditions set forth by the *Water Emergency Board* based upon the severity and anticipated duration of the shortage. Livestock watering will be permitted only on an as needed basis with a prohibition against nonessential use.
4. Filling, refilling, or adding of water to ponds and artificial lakes, swimming pools, spas will be prohibited.
5. Watering of all football fields, golf courses, and other playgrounds will be prohibited. Watering of parks, school grounds, and recreation fields will also be prohibited with the exception of plant materials classified to be rare, exceptionally valuable, or essential to the well-being of rare animals.
6. The use of water from fire hydrants shall be limited to fire fighting or to help maintain health, safety, and welfare of the public.
7. The operation of any ornamental fountain or similar structure is prohibited.
8. Permits for unmetered water services will not be issued or continued; construction water shall not be used for earthwork or road construction purposes.
9. The use of water for commercial manufacturing or processing purposes shall be permitted under conditions set forth by the *Water Emergency Board* based upon the severity and anticipated duration of the shortage.

4. Security Provisions Related to Malicious Acts

When the water supply system is well designed and manpower well trained, malicious actions can be prevented and their results contained. A water system owner or operator must perform an assessment of the *vulnerability* of its system to a terrorist attack or other intentional acts that are intended to substantially disrupt the ability of the system to provide a safe and reliable supply of drinking water.

A water system owner or operator must determine possible vulnerable systems, subsystems, and components, and identify those security measures that must be employed to protect the system and the customers or public it serves.

This "vulnerability assessment" (Bib2 app.2) is the identification of weaknesses in water system security, focusing on defined threats that could compromise its ability to meet its various service missions mentioned previously in the preamble; e.g. providing adequate drinking water, water for firefighting, and/or water for various commercial and industrial purposes. The assessment generally should address six basic elements and is performance based, meaning that it evaluates the risk to the water system based on the effectiveness and performance of existing and planned measures to counteract potential malicious actions. The "vulnerability assessment" elements are:

1. Characterization of the water system, including its mission and objectives.
2. Identification and prioritization of adverse consequences to be avoided.
3. Determination of critical assets that might be subject to malevolent acts that could result in undesired consequences.
4. Assessment of the likelihood (qualitative probability) of such malevolent acts from adversaries (e.g. terrorists, vandals).
5. Evaluation of existing countermeasures.
6. Analysis of current risk and development of a prioritized plan for risk reduction.

The *vulnerability assessment* addresses the following components of the water system: "pipes and constructed conveyances, physical barriers, water collection, pretreatment, treatment, storage and distribution facilities, electronic, computer or other automated control and monitoring systems which are utilized by the public water system, the use, storage, or handling of various chemicals, and the operation and maintenance of such systems." It is recommended to evaluate system vulnerabilities in cooperation with design, system operation, and the relevant level of security officer.

An emergency contact list must be prepared and distributed to the relevant personnel that will take part in a water emergency situation. Emergency response plans must be prepared, reviewed, and drill tested at appropriate intervals.

The system operator will have to identify

1. Critical services as well as critical customers, such as hospitals, emergency facilities, power facilities, central bus stations, and train stations, as well as other critical areas of the water system that if attacked could result in a significant disruption of vital community services and result in a threat to public health, or a complete shutdown of the system (e.g. inability to provide an adequate supply of water for fire prevention, inability to provide safe potable water, or release of hazardous chemicals that could cause catastrophic results).
2. Critical subsystems, facilities, services, and even singular locations in the system that if demolished or tampered with could result in significant disruption of vital community services or cause health protection hazards.

All assets that are considered critical can then be divided into categories that will identify their vulnerability to sabotage or destruction, e.g. high (H), medium (M), or low (L).

The system operator should contact the appropriate law enforcement officer to acquire information indicating the types of threats that may be likely against their facilities.

Some of the typical threats to the water systems may be relatively mild, e.g. vandalism, an insider sabotage caused by a disgruntled employee, others may be very serious, such as a planned external terrorist attack, or even a terrorist attack coordinated with a system employee causing a major system disruption.

An example of relevant questions related to the evaluation of water system vulnerability is as follows:

1. Do you have a written emergency response plan?
2. Is access to the critical components of the water system (i.e. a part of the physical infrastructure of the system that is essential for water flow and/or water quality) restricted to authorized personnel only?
3. Are all critical facilities fenced, including well houses and pump pits, and are gates locked where appropriate?
4. Are all critical doors, windows, and other points of entry such as tank, roof hatches, and vents kept closed and locked?
5. Are warning signs (tampering, unauthorized access, etc.) posted on all critical components of your water system? (e.g. well houses and storage tanks.)
6. Is there external lighting around all critical components of your water system?
7. Do you patrol and inspect all source intakes, buildings, storage tanks, equipments, and other critical components?
8. Is the area around all the critical components of your water system free of objects that may be used for breaking and entering?
9. Do you have an alarm system that will detect unauthorized entry or attempted entry at all critical components?
10. Are entry codes and keys limited to water system personnel only?
11. Do you have an updated operations and maintenance manual that includes evaluations of security systems?
12. Are your wellheads sealed properly?
13. Are observation/test and abandoned wells properly secured to prevent tampering?
14. Are well vents and caps screened and securely attached?
15. Are deliveries of chemicals and other supplies made in the presence of water system personnel?
16. Are chemicals, particularly those that are potentially hazardous (e.g. chlorine gas) or flammable, properly stored in a secure area?
17. Are vents and overflow pipes properly protected with screens and/or grates? Are tank ladders, access hatches, and entry points secured?
18. Can you isolate your storage tank from the rest of the system?
19. Do you control the use of hydrants and valves?
20. Has your system implemented a backflow prevention program?
21. Have water system personnel been advised to report security vulnerability concerns and to report suspicious activity?

An emergency response plan must be prepared to cope with the consequences of a malicious act, The emergency response plan will incorporate the results of the *vulnerability assessment* as completed for the system and will includes plans, procedures, and identification of equipment that can be implemented or utilized in the event of a terrorist or other intentional attack on the water system.

The emergency response plan will include actions, procedures, and identification of equipment that can obviate or significantly lessen the impact of terrorist attacks or other intentional actions on the public health and the safety, and supply of drinking water provided to communities and individuals.

5. Necessary Developments and Some Research Work

1. Develop an integrated methodology for assessing critical infrastructure.
2. Determine an agreed methodology for defining and assessing security risks at water utilities.
3. Develop a training package for water utility supervisors and managers, and water consultants, on systematic risk assessment methodologies.
4. Develop a system for continuous monitoring of certain defined contaminants.
5. Develop a proven technique to detect chemical and biological agents in water.
6. Develop methods to quickly determine contaminant introduction locations.
7. Develop a training package for water utility personnel based on vulnerability assessment results and developed methodologies and devices.

6. Concluding Remarks

Most of the present operating water systems were not built with the aim of coping with terror or other security threats. As a result, those systems are now vulnerable to the dangers imposed by the present situation embodied by the various threats as identified above. The recommended vulnerability assessment, security provisions, and related actions will reduce the vulnerability of and security threats to the water supply systems and will enable water supply system operators recover from an emergency situation through an approved and a well-rehearsed sequence of operation.

References

1. Gleick, P.H., IWRA, M. (1996) Basic Water Requirements for Human Activities: Meeting Basic Needs. Pacific Institute for Studies in Development, Environment, and Security, 1204 Preservation Park Way, Oakland, CA 94612, USA.
2. Office of Water EPA 810-B-02-001 (2003) Instructions to Assist Community Water Systems in Complying with the Public Health Security and Bioterrorism Preparedness and Response Act of 2002, January 2003.
3. Consider the Source: A Pocket Guide to Protecting Your Drinking Water, Office of Ground Water and Drinking Water, June 2002.
4. Anthony Allan, J. Water Security Policies and Global Systems for Water-Scarce Regions , The World Bank.
5. Water Security: A Preliminary Assessment of Policy Progress since Rio (2001) Prepared by the World Water Assessment Programme as a contribution to the International Conference on Freshwater (Bonn, December 2001) and the World Water Development Report.
6. Mason, S.A. (2004) From Conflict to Cooperation in the Nile Basin Interaction between Water Availability, Water Management in Egypt and Sudan, and International Relations in the Eastern Nile Basin. Conflict Sensitive Interviewing and Dialogue Workshop Methodology, Center for Security Studies, ETH Zurich.
7. Margat, J. and Vallee, D. (1999) Mediterranean Vision on water, population and the environment for the XXIst century, Contribution to the World Water Vision of the World Water Council and the Global Water Partnership prepared by the Blue Plan in the framework of the MEDTAC/GWP, December 1999.

1.3 WATER QUALITY SECURITY MANAGEMENT

JACOBO SACK
Private consultant to the Israel Water Commission, Tel Aviv, Israel

Abstract: In order to be able to cope with emergency situations regarding water quality, a water distribution system, should take measures for the prevention of the occurrence of water emergency events, control measures in case prevention measures failed, and mitigation measures in order to reduce the damage to the public served by the water distribution system if such an event should occur.

Keywords: water security, water emergency, prevention measures, control measures, water quality

1. For Prevention Purposes

1.1. DESIGN A NEW SYSTEM OR MAKE A SURVEY OF AN EXISTING SYSTEM TAKING INTO ACCOUNT WATER SECURITY ISSUES

At the design stage, security considerations should be taken into account and planned for, such as:

- Physical barriers to deter intentional attacks to different parts of the water distribution system; preferentially, protection standards should be issued for each component of the water distribution system, e.g. reservoirs, wells, intakes, pipelines, pumping stations, and water treatment plants.
- Installation of electronics elements to detect intrusions of unauthorized elements to the different sites of the water system.
- Alarms and alarm levels to be transmitted to control rooms.
- Software programs for adequate responses to alarms.
- Protocols for water emergency responses and ways to deal with a water security event.

1.2. DETERMINE THE BASE THREATS

Determining the base threats is a very difficult but necessary task, in order to define what you are preparing for. Therefore, this is an essential task that should be dealt with the help of multidisciplinary experts like water quality engineers, chemists, and intelligence people. There are three main categories of threats that should be dealt with: chemical, biological, and radioactivity threats.

Lists of the most probable threats in each category should be drawn. In order to prepare these lists, the criteria that should be considered include: availability of the material, solubility, toxicity, means and ways of attack to be considered, etc., such as described in Burrows and Renner (1999); Interim Voluntary Guidelines for Designing an Online Contaminant Monitoring System, ASCE (2004); in "The Threat to Water Supply Systems" by Sharan; etc.

Definition of the threat potential, allows to determine the volumes of water threatened that should be protected, as well as a criteria for prioritizing protection actions to be taken regarding the different parts of the water system.

Y. Sharan et al. (eds.), Water Supply in Emergency Situations, 17–20.

1.3 CREATE, CONTACT OR JOIN AN INFORMATION CENTER THAT COLLECTS RELEVANT INFORMATION FROM ALL AGENCIES INVOLVED AND TRANSFERS THIS INFORMATION TO THE AGENCIES CONCERNED

Since the information needed to deal with a water event is multidisciplinary, and involves many agencies, it is necessary to establish a center where all information is collected and managed.

This information concerns:

1. Standards for protection of works
2. Protocols for alarms
3. Threats and their characteristics
4. Protocols for first responders
5. Protocols for emergency event management
6. Protocols for cleaning a damaged water system
7. Public information protocols
8. Information distribution and handling protocols
9. Public security handling
10. Hospitals and health services during a water emergency event
11. Others

This information is essential during the preparation, control, and mitigation steps, and should be managed in such a way that it should be readily available to the people involved in the different aspects of water security management.

1.4. PREPARE "UMBRELLA" PROCEDURES THAT ARE UNIVERSAL FOR ALL AGENCIES INVOLVED

Since managing an event involves the coordination of several independent agencies, there is a need to prepare united protocols, prepared and accepted by all agencies involved (water distribution companies, local government agencies, health agencies, police, security agencies, social workers agencies, public information agencies, etc.).

1.5. IMPLEMENT PHYSICAL BARRIERS AND HARDEN WATER SUPPLY SYSTEMS THAT HINDER POSSIBLE ATTACKS

Build the physical barriers according to the standards designed in paragraph 1.

- Survey periodically the wholeness of the barriers and of the hardening devices that will be installed in the different parts of the distribution system.
- Update electronic devices for intrusion detection and alarm transmission.
- Update software for alarm handling and response at the control room level.

2. For Control Purposes

2.1. PREPARE A COMPREHENSIVE ONLINE SYSTEM THAT INCORPORATES CONVENTIONAL AND BIOASSAY ANALYSES

In the market, there are a few online monitoring instruments that might be helpful in detecting changes in water quality due to water security events. As well, there are commercial online bioassay analysis designed to detect toxicity of the water.

Since these systems are costly, a study of the system should be carried out in order to decide where and which systems should be installed, based on a cost-effective approach, as described in the overview on "Technologies and Methods for Rapid Detection of Terror Events" by Negreanu.

2.2. PREPARE PROCEDURES THAT ALLOW PROMPT DETECTION, RESPONSE, AND PUBLIC NOTIFICATION

Adequate procedures, correctly applied, can shorten the time of contamination exposure, therefore limiting the extent of the physical and health damage that can be inflicted upon the served population.

Therefore, procedures should be prepared for alarm levels activation, and immediate response plans, such as water sampling and test at the site and in the laboratories and rapid closure of water supply in the section of the water supply system affected.

Ready procedures for public notification can diminish the degree of panic and fear that can result from a water emergency event.

2.3. PREPARE LABORATORIES FOR A CAPACITY TO DETECT THE VARIOUS CONTAMINANTS

Once the threat has been identified, laboratories have to be prepared to be able to analyze the presence of the different potential contaminants in the water.

It is convenient to have more than one laboratory with the capacity to analyze the threats and that laboratories should be distributed geographically in order to minimize the time of response.

2.4. PREPARE STANDARDS FOR SAMPLING AND ANALYTICAL PROCEDURES (FIELD TEST KITS) FOR SAMPLING TEAMS

For the sake of prompt reaction, it is desirable to have sampling teams geographically distributed. The sampling procedures should be identical for all these teams in order to get reliable results of the analysis. These sampling teams should also be able to carry out the field tests that can be applied by the use of test kits on the site.

2.5. ROUTINELY EXERCISE THESE PROCEDURES

A plan for routine exercising of all the relevant procedures mentioned above should be drawn and put into effect several times a year.

Top table drills, field drills, and drills at all levels should be implemented, involving all agencies, laboratories, field teams, etc. in order to maintain a fit and able response crew. Lessons from these exercises and drills should be learned and applied continuously.

3. For Mitigation Purposes

3.1. AFTER INITIAL DETECTION, RESPOND *IMMEDIATELY* WHILE VERIFYING AFTERWARDS

Water velocity in a water supply system is about 1 m^3\s. Therefore, time is a primordial factor in contamination spread along the system. In order to minimize the damage caused by contamination of water, it is essential to act with maximum speed to achieve minimum spread of the contaminant and minimum of consumers that are exposed to contaminated water.

Therefore, after initial detection of contamination indications, there is a need for immediate response that can be to shut down the supply from the suspected contaminated water, by closing parts of the system or by changing operation of the water distribution system in such a way that the suspected water should be directed away from consumers. Verification should come only after the above-described initial response has been taken.

3.2. PREPARE ALTERNATIVE METHODS FOR WATER SUPPLY AND TECHNIQUES FOR WATER SUPPLY RENOVATION AFTER SHUTDOWN

When water supply is shut down to a part of the supply system, it is very important to renew the supply to the consumers as soon as possible. Therefore, plans should be ready for the supply of alternative water to the consumers. Preferentially, when it is possible, it is convenient to have alternative water sources that can be conveyed through the supply system. If no alternative sources exist, then means should be prepared for distribution of water at central distribution taps and/or for the distribution of water by tankers, water bottles, and any other possible mean.

3.3. PREPARE PROCEDURES FOR CLEANING OF WATER SUPPLY SYSTEMS, AND PREPARE THE NECESSARY EQUIPMENT

Cleaning of a contaminated water supply is probably the most complex operation when dealing with a water emergency situation that requires cleaning of the system. It is necessary to define *How clean is clean* with reference to the list of threats. Procedures and methodology should be prepared for the cleaning of a water system. Teams should be trained in the procedures for the cleaning of the system. Equipment and materials should be ready to be used by the necessary teams.

3.4. ROUTINELY EXERCISE THESE PROCEDURES

The only way to assure that the procedures will be put into effect correctly when the situation arrives is by exercising the people involved in the procedures application. Therefore, a routine plan for carrying out exercises at all levels should be implemented. Lessons from these exercises and drills should be learned and applied continuously.

4. Conclusion

In view of the threats posed by terror attacks, natural disasters and operational mishaps to drinking water distribution systems, water utilities should plan in advance and implement those plans on the ways to cope with water distribution interruptions or contamination, due to one of those threats taking place.

Plans and preparations should regard the means for prevention against the threats, means for control of the events if they happen and for the mitigation of the damage that can be caused by water emergency event occurrence to the public served by the water distribution system if such an event should occur.

References

1. Burrows, W.D. and Renner, S.E. (1999) Biological warfare agents as threats to potable water. Environmental Health Perspectives 107 (12):975–984.
2. American Society of Civil Engineers (ASCE) (2004) Interim Voluntary Guidelines for Designing an Online Contaminant Monitoring System, December 9, 2004.

1.4 ORGANIZATIONAL ASPECTS OF WATER SUPPLY

H. COCCOSSIS,[1] M. THEOCHARI, K. DIMITRIOU, AND G. KALLIS
University of Thessaly, Greece

Abstract: Our era is characterized to a large extent by uncertainty and unprecedented fluidity. This manifests the dynamics of modern society, a society where great changes take place, not only in the natural environment, but also in the socioeconomic and technological fields. Change leads to pressures on some already stressed natural resources, therefore resulting to a vital need of safeguarding them for a sustainable future. Water resources worldwide are increasingly important, facing significant threats and risks.

Water resource security is one of the most sensitive issues, especially in the water-scarce area of the Eastern Mediterranean. The main threats putting water in jeopardy include environmental pollution and contamination, overconsumption of available reserves, droughts, floods, and other natural hazards, which affect both quality and quantity of available resources. To these, one can add technological hazards, such as power loss of processing units, operational failure, or even intentional sabotage. The advances in technology and forecasting, the establishment of appropriate regulation and innovative measures, have greatly contributed towards the protection of resources, while being important for strategic management.

Sustainable development is a long-term process to achieve multiple social, economic, and environmental goals, and by definition, presupposes the safeguarding of key resources. The whole process is widely participatory and entails raising awareness and sensitization of all the different actors involved in it. It is important to identify the ensuing synergies within which, each actor has its own function. The public and private sectors, together with nongovernmental organizations (NGOs) need to have clear and definite roles within the framework of sustainable development. Delineating the role and the appropriate commitment of each of the actors not only prevent but also greatly help in responding effectively in emergency situations.

Keywords: urban ecosystem, organizational aspects, risk management, integrated water management, sustainable development, natural hazards, technological hazards

1. Introduction

Modern societies face great changes, not only in terms of the natural environment, but also in the socioeconomic and technological fields. Change leads to pressures on some already stressed human and natural ecosystems, therefore resulting to a vital need of safeguarding the latter against arising threats and risks.

The main threats putting water into jeopardy include environmental pollution and contamination, over-consumption of available reserves, droughts, floods, and other natural hazards, which affect both quality and quantity of resources. To these, one can add technological hazards, such as power loss of processing units and operational failure, and even intentional sabotage. The advances in technology and forecasting, the establishment of appropriate regulation and innovative measures, have greatly contributed towards safeguarding of resources, while being important for strategic management.

---

[1] University of Thessaly, Department of Planning and Regional Development, Environment and Spatial Planning Laboratory.

Y. Sharan et al. (eds.), Water Supply in Emergency Situations, 21–26.

Water resources security is one of the most sensitive issues, especially in the water-scarce area of the Eastern Mediterranean, which faces rising demand due to the growing urban population and the seasonal influx of visitors in tourist areas. To some extent, this increase in demand is connected to the modern way of living (which adopts increasing consumption patterns), urbanization patterns and tourism development. Mounting demand for clean water combined with the ever-increasing irregularity of weather phenomena and long-lasting dry spells, affect the entire region.

Another recent characteristic of the Mediterranean region is advancing urbanization, which reduces vegetation and increases built-up surfaces leading to lower water retention and infiltration rates and reduced replenishment of local groundwater reserves. However, the overall impact of urbanization on the quantity of groundwater reserves may well be positive due to increased infiltration from urban irrigation and due to seepage from distribution network losses. Urban sprawl has aggravated the problem of the redundant or poorly planned water system infrastructure, and has intensified problems like leaks in the water network. In addition, many areas lack proper land use and physical planning leading urbanization of areas subject to inundation, increasing the risks to human life and property.

Urban aquifers and streams are often irreversibly polluted in several Mediterranean cities. Pollution increases pressures on existing and potential freshwater sources. Infiltration of chemicals and other pollutants from the surface runoff (e.g. oil residues from cars) and the diffused pollution from landfills or household sewage pits contaminate aquifers. Agriculture, industry, and upstream cities may also contribute to pollution of the drinking sources.

Seasonality of both supply and demand due to tourism, a particular feature of the Mediterranean region, aggravate water resource problems. During the summer months, water shortages become more frequent. Distribution networks left unused during the winter period face overload pressures in the summer. On the other hand, designing the system with excess capability to satisfy tourism-related summer peak demands raises construction and maintenance costs significantly.

In cities where groundwater is utilized, overexploitation of groundwater reserves is an important issue. Coastal urban areas are prone to saltwater intrusion to the aquifer and salinization of groundwater reserves. Furthermore, such cities are more vulnerable to floods because they are located at the downstream outlet of basins and receive upstream runoff. Studies have shown that runoff in urban areas is 1.1–4.6 times greater than pre-urban runoff (Ameziane, 2002).

2. Water Supply System

The failure to incorporate the technological advances and organizational improvements into the traditional water supply systems has left the system unprotected against the rising threats of the ever-increasing droughts, floods and flash floods, and other natural disasters. Furthermore, outdated technologies are hazard-prone to the water system, whether they relate to power support systems or chemicals (i.e. contaminants) and are prone to accidents. This situation is further accentuated by the uncertainty and unprecedented fluidity that characterizes our era.

The urban water system constitutes a cross section of three broader systems: the river basin system; the coastal system; and the urban (eco)systems. Certain elements of these systems are within the borders of the urban basin, whereas others extent far beyond. The urban water system includes natural, modified, and man-made elements and processes, which produce both water services for an urban area and ecological functions in the broader basin and coastal zone. The typical definition of the water system in literature stops short in the description of the infrastructure components. However, if the urban catchment is perceived as an ecosystem, then within it, the urban water system spatial boundaries should extend to the limits of the urban basin or as far as the extent of influence of the urban center. Within the urban ecosystem, water follows a modified hydrological cycle, the urban water cycle.

The urban water cycle is one among the many metabolizing processes that constitute the ecosystem of the urban basin. The natural elements of the urban water system, i.e. the watercourses, wetlands, groundwater, urban canals, lakes and other water bodies (and the coastal waters in the cases of a coastal area), ensure health, balance, and replenishment of the system, provide water for human consumption, green spaces, and woodland, and offer assimilation capacity for waste (Hengeveld and de Vocht, 1982). Natural ecosystem processes also contribute to many of the desired services such as flow stability, purification, habitat provision and amenity, and aesthetic pleasure (PCE, 2000).

The structural elements of the urban water system aim to adjust the natural processes and inputs to the requirements and needs of the urban ecosystem. They include location subsystems, i.e. the physical entities in which water is changed in quantity and quality or used consumptively (reservoirs, treatment plants) and transfer subsystems that connect or feed the location systems (urban hydrology, pipes, sewers) (Hengeveld and de Vocht, 1982). They can also be classified according to operation into three general subsystems: water supply; wastewater collection, treatment and disposal; and drainage and flood protection.

As the urban ecosystem changes and evolves, management of the water supply system through specialized infrastructures and regulatory mechanisms becomes imperative to accommodate change.

There is a growing need for an integrated management of urban water systems that would take into account all aspects of the system and aim to minimize inputs and maximize the desired outputs while minimizing also the undesirable outputs. As undesirable outputs could be characterized any failures in managing natural processes and human modifications that could lead to contaminated drinking water, floods damaging urban infrastructure or pollution contaminating receiving waters and coastal wetlands.

From a broader respective, the consumption of energy in urban water operations contributes to greenhouse gases emissions. Wastewater treatment produces sewage sludge that must be disposed on land. The construction of reservoirs often has environmental and social impacts (e.g. displacement of populations and habitat alteration by big dams). Use of water for the urban area may impact on rival use(r)s of the same sources (e.g. other settlements, agriculture, environmental preservation, etc) (UNEP/MAP/PAP, 2004).

3. Organizational Aspects

Sustainable development is a long-term process seeing to achieve multiple social, economic, and environmental goals and by definition, presupposes safeguarding of critical resources.

The whole process is widely participatory and entails awareness raising and sensitization of all the different actors involved in it. It is important to identify the ensuing synergies within which, each actor has its own function. The public and the private sectors, together with the nongovernmental organizations (NGOs) need to have clear and definite roles within the framework of sustainable development. The delineation of the role and the appropriate appointment of each of the actors are essential to prevent and also to respond effectively to emergency situations.

The water supply system is characterized by intrinsic complexity not only regarding the spatial complexity and the different levels of operation (national or regional level, urban, municipal, etc.), but also regarding the multiplicity of different actors that influence and support the organization and operation of the system. This variability broadens even more when considering water supply systems in different countries. There is a wide array of organizational structures in water services ranging from public to mixed to fully private services.

Moreover, there are differentiations regarding the different water services and parts of the water supply system. The fragmented nature of the existing organizational schemes does not encourage integration and a holistic view of the services, where social equity, service quality, economic efficiency, and environmental protection are upheld. There is significant need for reform, however, there is not, and there should not be, one organizational model to implement on all water systems (UNEP/MAP/PAP, 2004). The success of an organizational arrangement depends on the circumstances and local context.

It is rather usual for the water utilities to alternate from time to time between public and private, and more often than not, combine the two. The main issue is that the performance of the arrangement remains high in terms of criteria such as integration, effectiveness in achieving goals, financial viability, and democratic accountability (UNEP/MAP/PAP, 2004). Flexibility very often enhances the ability of achieving high-standard services.

Any organizational change should result from an in-depth prolific study of scientific and technological advances, and the adoption of a participatory process which would involve all affected stakeholders. The whole process has to be open, transparent and inclusive (Gleick et al., 2002), based on equity values and scientific knowledge, since the quality of the process is in itself a target of organizational change (UNEP/MAP/PAP, 2004). The various features of a proposed organization change should be carefully studied before a decision is taken. Learning from experience and adoption of best practices is important. Finally, different actor modeling is difficult and the choice of a suitable model is made in conjunction with the existing political and social circumstances. The costs from social reaction should be taken into account when deciding on an organizational change. Scenario development, monitoring of interactions, and flexibility to adjust are also of crucial importance.

4. Risk Management

Over the past few years, the population increase in urban centers, economic growth and prosperity, and intensified resource use have added to the vulnerability of the urban centers and the Mediterranean in particular to climatic changes (Ameziane, 2002; Attané and Courbage, 2001). The water supply system is a fundamental part of the urban ecosystem, and one that is fairly exposed to natural and technological hazards.

Natural hazards are defined as natural, geological or meteorological phenomena that (Mearns and Overmars, 2000), can cause damage to water system infrastructure and put at risk the availability of water resources in terms of both quantity and quality. Accordingly, technological hazards result mainly from failure in the structural infrastructure, such as power loss or fire of water processing units and contamination. It is not uncommon for natural hazards to aggravate existing technological hazards, i.e. another, though rare, hazard that should be taken into account is intended sabotage. Emergency situations resulting from hazards have impacts on economic, environmental, or social sectors, and can affect more than one sector at a time.

The probability and severity of a hazard together with the vulnerability of the system, give us an indication of the risk involved. Risk has technical, scientific, and social dimensions, as it depends also on human activities and public perception on what is risk and to what point is it acceptable (Rees, 2002). Risk management is a vital process (or processes) which aims to manage and reduce the likelihood of any crisis arising from various hazards. It can be divided into two parts: first has a proactive nature and aims to assess and control risk so to avert any crisis; second part is mainly responsive, and involves the preparation in the case of an emergency and the contingency planning.

Risk planning, first of all requires a risk analysis based on a quantitative or qualitative assessment of the probability of harm of a given inherent hazard. Risk analysis defines the impacts and acceptable levels of risk (WDCC, 1998; Harrop and Nixon, 1999; EEA, 2004;). The impacts and hazards identified in a risk assessment have to be prioritized on the basis of the probability and severity of their consequences (Ale, 2002).

A wide array of tools can be used for risk planning, including disaster scenarios which help illustrate a situation and therefore arrive to response plans and procedures with the participation of all actors involved in an emergency situation (Mearns and Overmars, 2000). Public participation in planning is essential for the production of pragmatic risk plans, although it has to be carefully balanced with expert knowledge. Tools to assess the vulnerability include tree diagrams (WDCC, 1998), cause-and-effect or flow diagrams.

The next step in risk planning is the mitigation and reduction of a risk with technical or nontechnical measures, where possible. In all cases, monitoring, forecasting, and early warning systems are fundamental for preparedness and timely response. Furthermore, being prepared involves the little details as well, like access and sufficiency of materials and spare parts (Mearns and Overmars, 2000).

However, there will always be cases when, whatever the mitigation measures, crisis arise. In such a case, contingency responses may take effect. Careful planning at this stage is crucial for a contingency plan to define very clearly the role of each stakeholder and actor in the emergency situation. In the ensuing action list, the authorities, management, and employees are charged with specific responsibilities. Furthermore, all the different response phases (i.e. immediate, partial or temporary, and full service restoration) should be outlined. Although organization and planning may occur at the river basin or administrative level, the utility services have usually the responsibility for the emergency response (Suzenet et al., 2001). Finally, the plans should be updated and revised regularly in order to maintain flexibility and adaptability to technological and social changes.

5. Conclusions

It is apparent that there is an increasing call for integrated management of water systems that would maximize the desired outputs while minimizing the adverse ones. This should be done in the context of sustainable development and according to its specific objectives and mechanisms. The intrinsic complexity of the water supply system makes any organizational change difficult, time consuming, and with uncertain results. Sustainable development is also a long-term process that involves all affected stakeholders and demands a clear view of all spatial complexities, different levels of operation, and differentiations regarding the different water services and parts of the water supply system. Integrated water management within the framework of sustainable development, requires careful planning, including a contingency plan that will define very clearly the role of each stakeholder and actor in any emergency situation. All different response phases should be delineated. Emergency plans should always be kept updated and adjusted regularly, in order to follow the technological and social demands of their times.

References

1. Ale, B.J.M. (2002) Risk assessment practices in The Netherlands, Safety Science 40:105–126.
2. Ameziane El Hassani, T. (2002) Drought Preparedness and Risk Management in the Mediterranean Region, Paper presented at the IUNC Mediterranean Regional Roundtable, Athens, Greece, December 2002.

3. Attané, I. and Courbage, Y. (2001) La démographie en Méditerranée: Situation et Projections, Economica, Plan Bleu (Les Fascicules du Plan Bleu no. 11) Paris.
4. European Environmental Agency (EEA) (2004) Multilingual Environmental Glossary, available at: http://glossary.eea.eu.int/EEAGlossary
5. Harrop, D.O. and Nixon, J.A. (1999) Environmental Assessment in Practice, Routledge, New York.
6. Hengeveld, H. and de Vocht, C. (1982) Role of Water in Urban Ecology, Developments in Landscape Management and Urban Planning, vol 5, Elsevier, Amsterdam.
7. Mearns, A. and Overmars, M. (2000) Guidelines for Water and Sanitation Utilities, Risk Management Planning: A Disaster Preparedness Approach for Small Island Development States, SOPAC Technical Reports, Suva, Fiji Islands.
8. Parliamentary Commissioner for the Environment (PCE) (2000) Ageing Pipes and Murky Waters. Urban Water System Issues for the 21st Century, Parliamentary Commissioner for the Environment, Wellington.
9. Rees, J.A. (1998) Regulation and Private Participation in the Water and Sanitation Sector, TAC Background Papers No. 1, Technical Advisory Committee, Global Water Partnership, Stockholm.
10. Suzenet, G., Kallis, G. and Coccossis, H. (2001) Sustainable Management of Water for the City: Policy Guidelines, University of the Aegean, Mytilini, Greece.
11. UNEP/MAP/PAP (2004) 3rd Draft Guidelines for Integrated Coastal Urban Water System Planning And Management in Coastal Areas of the Mediterranean, Priority Actions Programme, Mediterranean Action Plan, Split.
12. Western Drought Coordination Council (WDCC) (1998) How to Reduce Drought Risk, Preparedness and Mitigation Working Group, Texas.

CHAPTER II

COUNTRIES' AND MAJOR CITIES' CASE STUDIES

2.1 ALTERNATE WATER SUPPLIES IN EMERGENCIES

SARAH BURN

London Resilience Team, Government Office for London, Riverwalk House, 157-161, Millbank, London, SW1P 4RR, England, e-mail: sburn.gol@go-regions.gsi.gov.uk

Abstract: Following the attacks on America on September 11, a comprehensive review was carried out to examine London's resilience and its ability to respond to, and recover from, a major or catastrophic attack.

After the completion of the review it was discovered that while London had good tried and tested procedures for dealing with "conventional" – e.g. car bomb – terrorism, which it had been handling for many years, a step change in planning and response capabilities was needed to confront the threat that London faced post September 11.

The *London Regional Resilience Forum* was set up in May 2002 to lead emergency planning. Led by Phil Woolas, the Minister for London Resilience as Chair and Ken Livingstone, the London Mayor as his Deputy, the Forum includes senior representatives of all London's frontline organizations such as the emergency services, central and local government, transport, utilities, national health service (NHS), military, and business.

Over the past 2 years, subcommittees and working groups have taken forward particular work streams and are supported by the multiagency *London Resilience Team* – a small core of civil servants who work with over 25 secondees from key organizations. This new structure integrates the efforts of all key frontline responders to help make London better prepared for a major or catastrophic incident.

Keywords: emergency planning, resilience, water supply legislation

1. The London Resilience

1.1. THE LONDON RESILIENCE PARTNERSHIP

The London Resilience Partnership (LRP) consists of stakeholders across London who have an interest in London's resilience.

These consist of

- Emergency services
- Heath service
- Voluntary sector
- Local, regional, and central government
- Utilities
- Military
- Transport
- Media
- Business

Y. Sharan et al. (eds.), Water Supply in Emergency Situations, 29–34.

1.2. THE LONDON REGIONAL RESILIENCE FORUM

The London Regional Resilience Forum includes representatives from across the LRP and is chaired by the Minister for London Resilience with Mayor Ken Livingston as his deputy. There are several reporting subcommittees that drive the work of the Forum forward.

1.3. THE LONDON RESILIENCE TEAM

The London Resilience Team is a multiagency group made up of a core team of civil servants and over 25 secondees from stakeholder organizations within the LRP.

Together they work toward:

- Coordinate emergency planning structure
- Provide support to government
- Ensure robust command, control, and coordination
- Coordinate review and exercise program
- Deliver capabilities program
- Emergency response support
- Special projects
- London resilience strategy

The team also serves as a secretariat to the LRRF and its subcommittees.

1.4. THE BUSINESS SUBCOMMITTEE

The business subcommittee brings together representatives of the business sector in London to provide a coordinated input into emergency contingency planning in the capital. It looks at how to encourage business continuity planning throughout the business sector and how to share relevant information throughout the community.

However the subcommittee is more than a means of highlighting the importance of resilience to the business community. The subcommittee aims to be a forum for business to express their views, which are fed back by the chair to the London Regional Resilience Forum.

The group has a wide membership encompassing organizations from all over the business community.

1.5. UTILITIES SUBCOMMITTEE

The Utilities Subcommittee (USC) harmonizes the key utility and associated organizations serving London. It aims to provide a coordinated input into emergency contingency planning for major incidents in London. The subcommittee also plays a key role in briefing the utilities sector in London on the importance of emergency contingency planning. The group also takes an overview of emergency contingency planning in the utilities sector.

The membership of the subcommittee includes the following organizations:

- Thames Water
- EDF Energy
- British Telecom
- National Grid Transco

- O2
- Telehouse Europe

As category 2 responders they are required to provide emergency services in their specialist areas along side the regular blue light services. They also sit on the London Regional Resilience Forum.

2. Alternate Water Supply Working Group

The Alternate Water Supply Working Group (AWS) has members from diverse backgrounds including military, business, health, police, and the water industry. They work to provide a structured process in the event of a failure in the normal water supply.

The water industry in London is owned by a variety of companies, the biggest of which is Thames Water. They represent all the London water companies at the USC and the AWS; they also manage all the wastewater needs for the region.

Through the USC, an understanding has been reached for all utilities to provide mutual aid, this would allow for the water company to call on the facilities available, including vehicles, land, equipment, and when needed, manpower.

There are over 30,000 km of mains within London and over 30% of these are over 150 years old, they serve 7.5 million people with 2,500 million liters of water a day. Detailed plans for the specific response to an event are being laid at this time, along with the cross sector cooperation that will give a solid response to any water-related need.

3. Civil Contingency Act

Following severe flooding and other events in the autumn and winter of 2000, the Deputy Prime Minister announced a review of emergency planning arrangements. The review included a public consultation exercise which reinforced the Government's conclusion that existing legislation no longer provided an adequate framework for modern civil protection efforts and that new legislation was needed.

3.1. OVERVIEW OF THE ACT

The Act, and accompanying regulations and nonlegislative measures, will deliver a single framework for civil protection in the UK to meet the challenges of the 21st century. The Act is separated into two substantive parts: local arrangements for civil protection (Part 1) and emergency powers (Part 2). The overall objective for both parts of the Act is to modernize outdated legislation.

Part 1 of the Act sets out the range of possible incidents which local responders must prepare for as set out in specific civil protection duties. Part 2 sets out the situations in which it may be possible to use emergency powers if appropriate safeguards are net.

3.1.1. Part 1: local arrangements for civil protection

The purpose of Part 1 of the Act is to establish a new statutory framework for civil protection at the local level. This, together with accompanying guidance and regulations, will set out clear expectations and responsibilities for frontline responders at the local level to ensure that they are prepared to deal effectively with the full range of emergencies from localized incidents through to catastrophic emergencies. It divides local responders into two categories.

Those in *Category 1* will have duties placed upon them to:

- Assess local risks and use this to inform emergency planning.
- Put in place emergency plans.
- Put in place business continuity management arrangements.
- Put in place arrangements to make information available to the public about civil protection matters and maintain arrangements to warn, inform, and advise the public in the event of an emergency.
- Share information with other local responders to enhance coordination.
- Cooperate with other local responders to enhance coordination and efficiency.
- Provide advice and assistance to businesses and voluntary organizations about business continuity management (local authorities only).

Those to be covered by the duties at present are:

Local authorities • All principal local authorities *Government agencies* • Environment agency • Scottish environment *Protection agency*	*Emergency services* • Police forces • British Transport Police • Police service of Northern Ireland • Fire authorities • Ambulance services	*NHS bodies* • Primary Care Trusts • Health Protection Agency • NHS Acute Trusts (Hospitals) • Foundation Trusts • Local Health Boards • Any Welsh NHS Trust which provides public health services • Maritime and Coast guard Agency • Health Boards • Port Health Authorities

Category 2 organizations will be placed under the lesser duties of cooperating with these organizations and sharing relevant information. Those to be included at present are:

Utilities • Electricity • Gas • Water and sewerage • Public communications providers (landlines and mobiles)	*Transport* • Network rail • Train operating companies (passenger and freight) • Transport for London • London underground • Airports • Harbours and ports • Highways agency	*Government* • Health and Safety Executive *Health* • The Common Services Agency (in Scotland)

The Act enables the minister to alter the membership of both categories of responders in order to ensure flexibility and to take account of future developments. The details of what this means in practical terms will be fleshed out in regulations and guidance. It is intended that category 1 and 2 organizations come together to form "Local Resilience Forums" (based on police areas) which will help coordination and cooperation between responders at the local level.

While it is primarily focused at civil emergencies, Part 1 will improve the UK's ability to deal with the consequences of a wide range of disruptions by improving the planning process at a local level, building better contacts between organizations and ensuring what goes on at the local level dovetails with efforts at the regional and national levels.

3.1.2. Part 2: emergency powers

In the UK, emergency powers allow the making of special temporary legislation to deal with the most serious of emergencies. They are not a means for instigating martial law, for undermining Parliament, banning political parties, or anything else of that nature. An essential point to note is that Emergency Powers legislation is a mechanism for dealing with only the most serious of emergencies that require an urgent response, an instrument of last resort. The previous emergency powers legislation (the Emergency Powers Act 1920) was used 12 times in its 84-year history, the last time being in 1974. Since then, a considerable amount of sector-specific emergency legislation has been introduced which reduced the need to resort to emergency powers, in part because of a recognition that Emergency Powers legislation was inadequate.

Nevertheless, there is still a need for a latent capacity to rapidly make new temporary statutory provision where this is the most effective way of enabling the resolution of an emergency situation. The government needs a tool that can be deployed to address all forms of disruptive challenge where existing legislation is insufficient.

The Act allows the making of temporary special legislation aimed at dealing with a serious emergency that fits within the definition. The Queen, as Head of State, will formally indicate that emergency powers are necessary as part of the Order in Council that makes the regulations themselves. For the first time a fallback option has been included to cover the possibility that emergency powers will be needed, where the Queen is, for whatever reason, unable to act. The Act therefore allows for a senior minister or the prime minister to make the regulations in the unlikely event that Her Majesty is not able to do so.

The Act introduces a range of other new features, mostly designed to ensure emergency powers cannot be misused and can be used in a more targeted and proportionate manner.

The centerpiece of these is the "triple lock," which ensures emergency powers will only be available if:

- An emergency that threatens serious damage to human welfare, the environment, or security has occurred, is occurring, or is about to occur.
- It is necessary to make provision urgently in order to resolve the emergency as existing powers are insufficient and it is not possible to bring forward a bill in the usual way because of the need to act urgently.
- Emergency regulations must be proportionate to the aspect or effect of the emergency they are directed at

In addition, emergency regulations:

- Cannot prohibit or enable the prohibition of participation in, or any activity in connection with, a strike or other industrial action.
- Cannot instigate any form of military conscription.
- Cannot alter any aspect of criminal procedures.
- Cannot create any new offence other than breach of the regulations themselves.
- Must be compatible with the Human Rights Act and EU law.
- Are open to challenge in the courts.

For the first time it is possible to use emergency powers on a regional and/or devolved administration basis. This ensures any special temporary legislation will apply only in the part of the UK affected by the emergency, leaving those elsewhere unaffected. The Act also requires the appointment of a "Regional Nominated Coordinator" ("Emergency Coordinator" in the devolved administrations). This individual will be a highly trained crisis manager with expert knowledge of the particular type of emergency in question who, if emergency powers are used, will act as the focal point for coordination of response efforts at the regional or devolved administration level.

As with the existing legislation, emergency regulations must be presented to Parliament for its approval as soon as practicable after being made. Parliament may amend the regulations and must approve them within 7 days of laying. If parliamentary approval is not forthcoming, the regulations cease to have effect. The maker of emergency regulations would be subject to an obligation to protect and restore the ability of Parliament to scrutinize emergency regulations, and the ability of the courts to entertain challenges.

The government has given a commitment to ask a senior Privy Councillor to conduct an inquiry within 1 year of any use of emergency powers. The report would be published and debated in both Houses of Parliament.

Acknowledgments

Thames Water Limited.
London Resilience Team – Government Office for London.
Civil Contingency Secretariat – Cabinet Office.

References

www.londonprepared.gov.uk
www.cabinetoffice.gov.uk/contingency_planning/
www.preparingforemergencies.gov.uk/
www.defra.gov.uk/environment/water/
www.ukresilience.info/
www.gos.gov.uk/gol/
www.ukresilience.info/ccact/index.htm

2.2 CONSEQUENCES OF A BIG FLOOD ON THE PARIS WATER SUPPLY SYSTEM

BRUNO NGUYEN*
9 rue Victor Schoelcher, 75014 Paris, France

Abstract: A major flooding of the Seine River would have tremendous consequences and would impact very significantly the daily life of the 10 million people living in the Parisian area. A deep study of the impacts of such a catastrophic natural hazard has recently been initiated by the French authorities. The study performed by all the actors involved water supply, power supply, transportation, telecommunication, waste water collection, and treatment and has revealed maps of risks and pointed out the districts of the city that would suffer interruption or degradation of the ordinary level of service. The operational difficulties and constraints associated with the big flood on the water supply have been estimated and backup solutions have been envisaged to face this crisis situation.

Keywords: drinking water supply, flood, Paris, risk analysis, risk management, emergency, alternate water supply

1. Introduction

The risk of flooding is the most common natural risk in France. Small and large rivers cross through most French towns and cities. For old cities, such as Paris, one can still find the signs of the high water mark from floods that struck the city over the course of the past centuries. The highest-level flood crest ever registered was in February 1658 when the Seine River reached a level of 8.81 m.

The flood during the winter of 1910 when the Seine reached a level of 8.62 m left numerous traces and was recorded thanks to the many photographs taken during the event. The estimated flow of the Seine during the 1910 flood was 2,400 m^3/s or 10 times its normal flow (Fig. 1).

The visible consequences of the 1910 flood are a good illustration of what could happen today if a climatic phenomenon of this scale were reproduced: in 1910, the Paris subway/underground transit system already existed and the city population was equal to today's population of 2,100,000.

Figure 1. Paris – The great flood of the Seine, January 1910

Y. Sharan et al. (eds.), Water Supply in Emergency Situations, 35–42.

The suburbs of Paris, on the other hand, have greatly expanded and developed since the beginning of the 20th century. The resulting damage caused by a big flood is thus more difficult to evaluate. Other factors complicating the evaluation of the aftermath of a major flood are the urban development along the embankments of the river and the impermeability of the terrain resulting from intense urbanization.

Since the year 2000, the local, regional, and governmental authorities have come to the realization that they must take into account the evolution of modern society in France: the citizens of today will no longer accept the risks of daily life and systematically search who is responsible when misfortune strikes.

The authorities have committed to a general review of the situation in the greater Paris region (Ile de France) so as to collectively prepare for the consequences of the next large flood of the Seine River. We do not know when it will happen, but we may be certain that it will occur sometime in the future.

2. The Risks Linked to a Major Flood Crest in Paris

The rise of the water level in the Seine during the last two major floods occurred slowly over several weeks which may explain their low number of fatalities: 50 deaths in 1658 and only one death in 1910.

The damage and destruction to buildings and infrastructure, and the resulting effect on economic activity were, however, of major proportions: the cost of the 1910 flood is estimated to have been €1.4 billion. A flood of such magnitude occurring today could possibly reach a total of €12 billion.

Dams have been constructed on the rivers upstream from Paris, but their capacity to stock water is only 830 million cubic meters, which would be insufficient when compared to the volume of 4 billion cubic meters of water produced by a big flood.

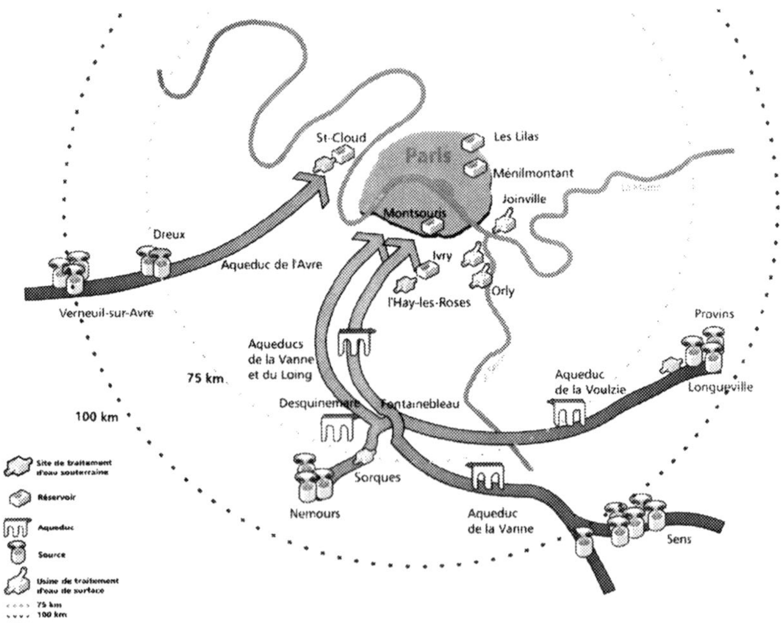

Figure 2. Origins of Paris Drinking Water Supply

The drinkable water supply system in Paris, as well as that of the sewer network, is still constrained by the decisions and orientations taken during the second half of the 19th century during the large public works projects realized under Napoleon III.

The network structure in 2005 is thus close to that which existed in 1910; two of the three water plants which treat river water and supply half of Paris with drinkable water existed in 1910. Water treatment technology has radically changed, but the production sites have remained the same. New reservoirs for potable water have been added, but the principles of distribution have not changed (Fig. 2).

The originality of the method chosen by the authorities was to ask each operator, infrastructure manager, or civil administration, to imagine the consequences of their installations of a rise in water level to that attained in 1910. The department of the environment for the Ile de France region created a CD-ROM with maps at a scale of 1/25,000 showing the different flood levels of sectors as they are progressively drowned under the floodwaters.

The maps are cut into little subfiles each one representing areas with a homogenous height on which the uncertainty of the water level that would be reached in the different scenarios is 15 cm.

The rise of the water level is broken up into nine scenarios starting from the moment that the waters overflow the riverbanks and continuing every 40 cm higher.

The map of "risks" presented in Fig. 3 makes visible the zones of Paris, which would be inundated at the surface and underground in the hypothesis of a flood crest at the level of 1910. We can note that the water spreads also using the basements and the subway/underground lines, reappearing in places in the form of isolated lakes.

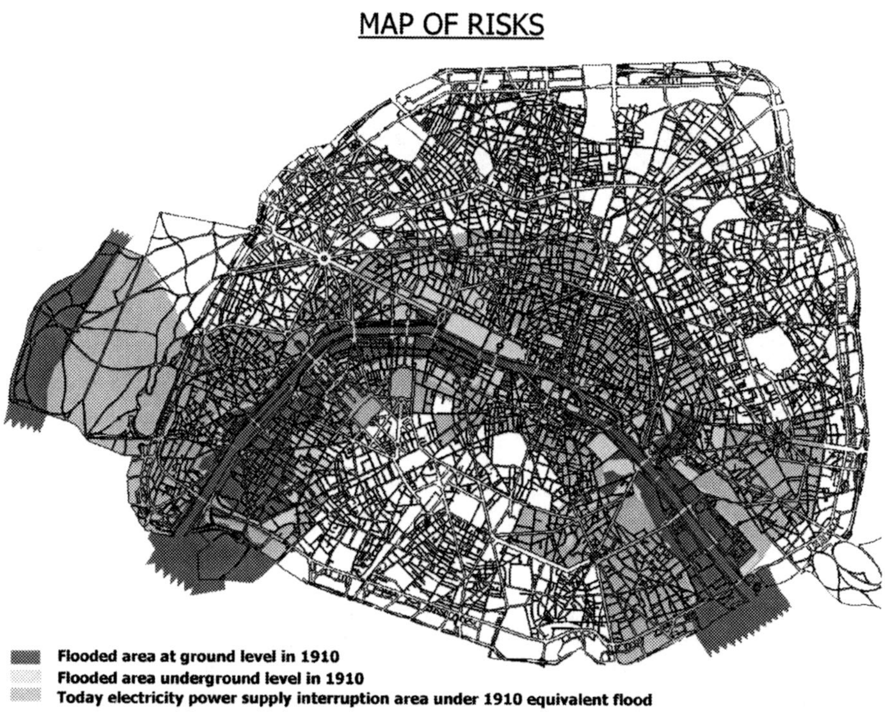

Figure 3. Map of risks

On the same map, one may also notice that the zone in which there will be an interruption of electrical current will much be larger than the zone of underground flooding.

The consequences in 2005 of a flood at the level of 1910 (Fig. 4) would be the following:

- Flooding of 20% of the surface of the city of Paris.
- Flooding of the National Assembly building and numerous ministries.
- Flooding of 21 embassies.
- 18 hospitals and clinics, 16 old-age homes, and 16 establishments for the handicapped.
- Three train stations (St Lazare, Lyon, and Austerlitz).
- Numerous metro stations and the Paris heliport.
- 29 important museums, 16 libraries.
- 384 historic buildings.
- 600,000 people will be in the flood zone; 270,000 of them living in Paris.
- 600,000 people will be without electricity in Paris.

On the basis of the maps and their subfiles and the scenarios, the authorities have asked the operators of utility, transport, and telecommunications networks to precisely estimate at what level and in what sectors their operations will be effected at different levels of the flooding.

Figure 4. A Street in Paris during the great flood, January 1910

3. The Supply of Potable Water Facing the Risk of Inundation

The consequences of the rivers overflowing their banks on the supply of water in Paris has been thoroughly studied following the different steps between the water taken from its natural environment to its arrival at the consumer's water tap (Fig. 5).

3.1. THE CONSEQUENCES ON PRODUCTION LEVEL

The average drinking water production in Paris is 615,000 m^3/day, but with the city partly under water, its public transport system largely blocked and the economic activity reduced, the volume of water needed on a daily basis would probably drop lower than 400,000 m^3/day.

Thus, the water production capacity would presumably be sufficient in quantity and quality in the case of a flood similar to that of 1910. However, the new European constraints with regards to drinking water quality have led to define and start the construction of new treatment processes for underground water. In the near future, by the end of 2007 (Fig. 6), these new treatment facilities (which will be beyond the reach of a flood at the level of 1910) will remove the risk of the rise of the water turbidity as a consequence of a big flood, but with a production capacity of underground water equally limited (200,000 m^3 for the South springs and 100,000 m^3 for the West springs).

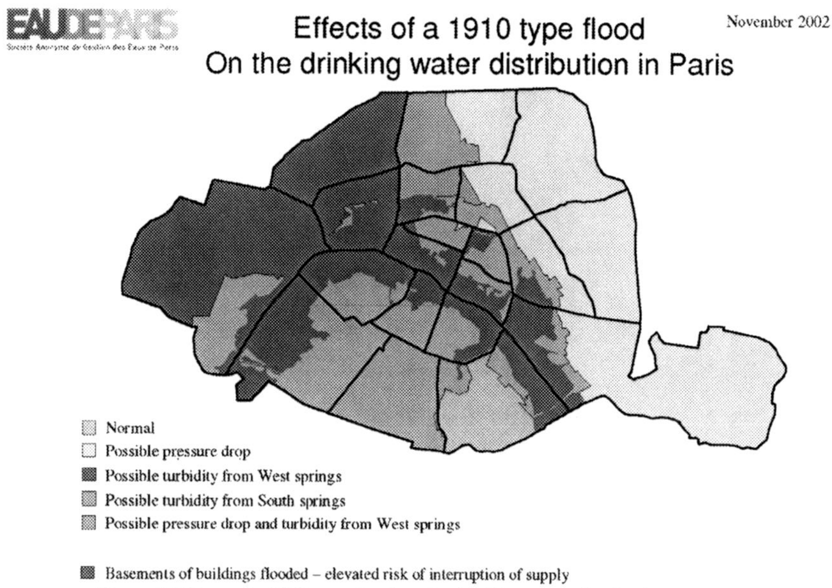

Figure 5. Effects of a 1910 type flood on the drinking water distribution in Paris

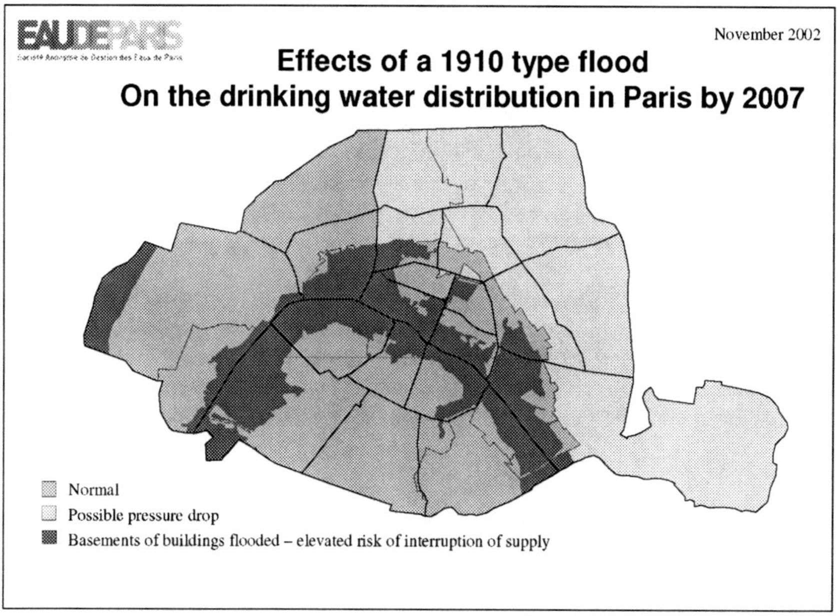

Figure 6. Effects of a 1910 type flood on the drinking water distribution in Paris by 2007

The production of potable water is divided as indicated in the following chart.

	Average production 2004 (m^3/day)	Maximum production (m^3/day)	Production in a flood like 1910 (m^3/day)	Comments
Ivry Plant	140,000	300,000	0	Plant under 1.88 m water
Joinville Plant	110,000	300,000	0	Plant under 0.57 m water
Orly Plant	110,000	300,000	200,000	Plant constructed after 1910, above water with a margin of +0.13 m electric supply using emergency generators, production capacity reduced by 2/3
South Springs	160,000	350,000	200,000	Risk or rising turbidity following probable heavy rain
West Springs	100,000	160,000	100,000	
Total	620,000	1,410,000	500,000	

3.2. THE CONSEQUENCES ON DISTRIBUTION

With regard to the distribution of water within the confines of the City of Paris, no reservoirs stocking potable water and no lift-pumping station would be touched; the latter are equipped with electric generators assuring them electric energy autonomy. On the other hand, the impact on the network lines may be classified in three categories, listed in descending order of importance:

- Impact on the remotely controlled valves situated in subterranean chambers; loss of remote control maneuvers.
- Impact on the electronic-measuring devices repatriated by the dispatching center; loss of information on water pressure, water quality, and flow.
- Impact on manual valves; impossible to perform maneuvers in the flooded galleries and sewers.

The loss of control of the remotely controlled valves is the most damaging because they involve a significant lowering of security norms (the reaction time would be much too long) and the impossibility of regulating water pressure. In such a case, it will be necessary to leave the remotely controlled valves in the open position.

A flood similar to that of 1910 would submerge 18 local stations (four of which are in plants) out of 150 and render them unusable for remote control. The number may be higher, however, if other local stations, which are not under water find themselves cutoff from the EDF and France Telecom networks (electricity and communications). It will therefore be necessary in case of mounting floodwaters to leave the network in a "standard" configuration.

4. Emergency Water Supply

There exists in Ile de France, from a depth of 600 m beneath the surface, a layer of very old water called the Albien (Fig. 7). This water resource was used in the 19th century through artesian wells that have since fallen into disuse. This source of very pure water has become the reserve for the supply of drinking water for emergency use and is prohibited for industrial use.

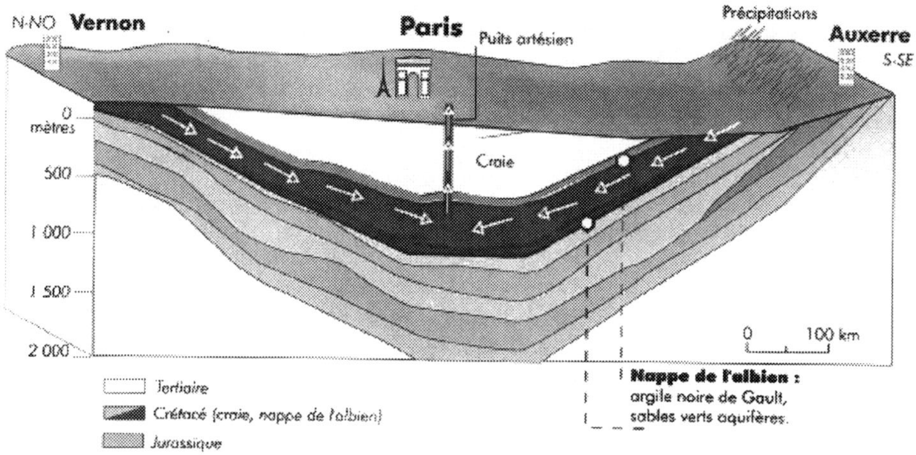

Figure 7. Water Table under Paris reserved for Emergency Supply

EAU DE PARIS has drilled five wells to reach the water in the Albien deposit in Paris (Fig. 8):

- The Madone Well in the 18th district in 1999.
- The Paul Verlaine Well in the 13th district in 1999.
- The Lamartine Well in the 16th district in 1996.
- The Henri Queuille Well in the 15th district in 2000.

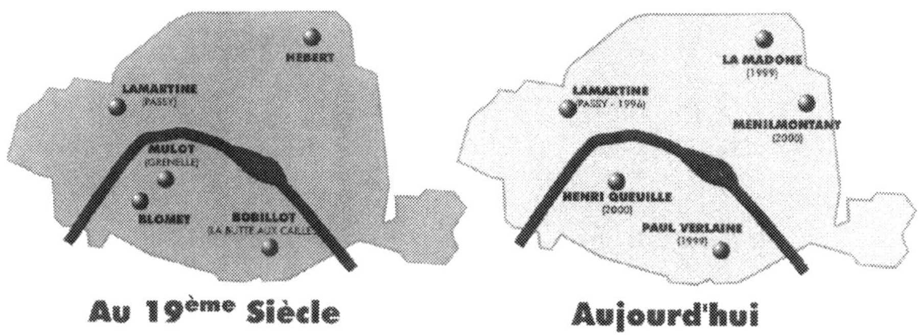

Figure 8. Wells drawing from the Albien deposit in the 19th century and today

Three of these wells are connected to a public fountain: Madone; Verlaine; and Lamartine. The water can be treated to remove excess iron before being made available to the public. The average flow for each fountain is 2 m^3/h. The citizens of Paris may use the water from the fountains at no charge.

The bottling factory in Auteuil uses the water pumped from the Lamartine Well at a pace of 10 m^3/h during its operation; on average 100 m^3/day. The water is produced in 5 gal (18.9 l) plastic bottles. This bottled water is marketed by a subcontractor who guarantees a rotating stock of 40,000 bottles immediately available to be distributed to the population in a crisis situation. The City of Paris made available this bottled water stock to the Indonesian government after the Tsunami.

In a crisis situation, the regions' supply plan previews the use of the emergency water supply for the Paris region. Each of the five wells is equipped with a reinforced production capability of 150 m^3/h for a total production of 3,600 m^3/day. The water pumped to the surface is made available at the well site. In an emergency situation the water will not be treated to remove excess iron.

It should be noted that in case of a flood at the same magnitude as that experienced in 1910, the five wells would be able to function.

The next step is the study for an alternate distribution system from the wells and the Albien water table to the 327 elementary schools of Paris. This second distribution network of pipes would be 156 km long and would require also permanent treatment for removal of excess iron and for disinfection. The estimated cost for this new distribution network is €30 million.

2.3 AN OVERVIEW OF WATER SUPPLY AND DISTRIBUTION SYSTEMS IN THE METROPOLITAN CITY OF ISTANBUL

ORHAN YENIGUN*
Institute of Environmental Sciences and Centre for Disaster Management, Bogazici University, 34342 Bebek, Istanbul, Turkey
BURAK DEMIREL
Institute of Environmental Sciences and Centre for Disaster Management, Bogazici University, 34342 Bebek, Istanbul, Turkey

Abstract: Istanbul is a large metropolitan city with a population over 10 million people. Water demand in Istanbul will be about 900 million cubic meter per year in 2010. Water management in Istanbul is a difficult task due to its topography and physical boundaries. Apart from normal management practices, management of risks of breakdown or damage to the distribution network due to a major earthquake is another task to be confronted by the water authorities in the city.

Keywords: Istanbul, water distribution, risk management

1. Introduction

Istanbul reached a population of around 10,000,000, which was almost 15% of the total population of Turkey in 2000. Increase in the population of Istanbul is 33.1‰ annually between 1990 and 2000, and the rate of net migration is 46.09‰ by 2000. Furthermore, Istanbul is located in the Marmara region, where the population increase is 26.7‰ per year between 1990 and 2000 (www.die.gov.tr/english). More recently, population in Istanbul was estimated to be 11,332,000 in 2005 (www.die.gov.tr). The Bosphorus bisects the city from north to south, acting as a physical boundary between the European and the Anatolian sides, and the total surface area and populations of both sides are roughly the same.

As the population growth in Istanbul seems a main issue nowadays, water demand is another closely associated matter. The water demand of Istanbul was computed to be 694 million cubic meter per year in 2000; it is presently about 800 million cubic meter per year and is predicted to be 913 million cubic meter per year by 2010 (www.iski.gov.tr). In addition to water demand associated with the population growth, management of water supply resources and the distribution network in case of emergency conditions, such as an earthquake or terrorist attack, are other significant issues for policy and decision makers. Therefore, management of water supply and distribution systems of the metropolitan city of Istanbul should be carried out accordingly.

2. Water Problem of Istanbul

A high increase in population is an important factor influencing the management of clean water resources in Istanbul. The population of Istanbul and the annual increase in population are reported to be 10.072.447 and 3.2%, respectively, in 2000 (Istanbul Water Sewage Administration Activity Report, 2003). As stated previously, the water demand of Istanbul was computed to be 694 million cubic meter per year in 2000, and is predicted to be 913 million cubic meter per year by 2010. Water demand of Istanbul is summarized in Table 1 (Istanbul Water Sewage Administration Activity Report, 2003).

Istanbul is located on both the European and the Anatolian sides, and water to the European side is supplied from Lake Terkos and three other water reservoirs. On the Asian side, there are four water reservoirs and associated distribution systems. There are two pipelines that traverse

Y. Sharan et al. (eds.), Water Supply in Emergency Situations, 43–48.

TABLE 1. Water demand of Istanbul

Year	Population (million)	Annual total water demand (million cubic meter/year)	Daily total water demand (million cubic meter/day)
1990	7.4	511	1.4
1995	8.4	657	1.8
2000	10.1	694	1.9
2010	12.1	913	2.5
2020	13.6	1.059	2.9
2030	14.6	1.241	3.4
2040	15.5	1.387	3.8

the Bosphorus at different points and connect the distribution network on both sides. Water resources and distribution systems of Istanbul are shown in Table 2, respectively (Istanbul Water Sewage Administration Activity Report, 2003). Total capacity of water resources of Istanbul is also displayed in Table 3 (Istanbul Water Sewage Administration Activity Report, 2004). Table 3 shows a substantial increase in the total capacity of water resources during the last decade in Istanbul gradually. In 1994, capacity of water resources was around 590 million cubic meter per year, while it was 1,170 million cubic meter per year during 2003. Almost a 100% increase can be observed in capacity of water resources of Istanbul between 1994 and 2003.

Istanbul Water Sewage Administration (ISKI), founded in 1981, operates under Istanbul Metropolitan Municipality, and is primarily responsible for management of water resources and water distribution network in Istanbul. The main tasks of ISKI are to provide drinking water, collection–treatment–disposal of wastewater, to protect water catchment areas, and to rehabilitate streams (Istanbul Water Sewage Administration Activity Report, 2004).

TABLE 2. A description of water resources of Istanbul

Water resources	Service year	Annual efficiency (million cubic meter/year)	Dry season average (million cubic meter/year)
Elmalı 1 and 2	1893–1950	15	12
Terkos	1883	162	80
Alibeyköy	1972	36	21
Ömerli	1972	235	164
Darlık	1989	97	72
Buyukcekmece	1989	120	45
Yeşilvadi Cevirme Y.	1992	10	6
Duzdere, Kuzudere, Buyukdere	1995	44	44
Şile Keson Kuyuları	1996	30	30
Elmalıdere, Sultanbahcedere, Kazandere	1997	131	131
Papucdere	1998	60	60
Sazlıdere Barajı	1998	85	55
Yeşilcay	2003	145	145
Total current annual average		1.170	865

TABLE 3. Capacity of water resources

Year	Capacity (million cubic meter/year)
1994	590
1995	649
1996	679
1997	860
1998	945
1999	945
2000	1.025
2001	1.025
2002	1.025
2003	1.170
2004	1.170

3. Water Distribution In Istanbul

Istanbul is located partially on both continents, and covers a total area of 5.712 km^2. Therefore, there exits a large water distribution network on both sides of the city. More detailed data about water distribution in Istanbul is also given in Table 4 (Istanbul Water Sewage Administration Activity Report, 2003) and in Fig. 1 (www.iski.gov.tr), respectively. Existing water treatment plants and the amount of water treated by these water plants in Istanbul during 2004 are summarized in Tables 5 and 6, respectively (Istanbul Water Sewage Administration Activity Report, 2003). A schematic of Kagıthane and İkitelli water treatment plants are also displayed in Figs. 2 and 3, respectively (Istanbul Water Sewage Administration Activity Report, 2004).

TABLE 4. Water distribution in Istanbul between 1884–2003

Water resources	Water distribution (million cubic meter/year)
Alibeyköy Dam	36
Elmalı 1 and Elmalı 2 Dams	15
Terkos Lake	162
Ömerli Dam	235
Darlık Dam	97
Yeşilvadi Regulator	10
Buyukcekmece Dam	120
Şile Keson Wells	30
Duzdere Dam	4.5
Kuzuludere Dam	11.3
Buyukdere Dam	28.4
Sazlıdere Dam	85
Elmalıdere Regulator	11.6
Sultanbahce Dam	19.4
Kazandere Dam	100
Papucdere Dam	60
Yeşilcay Regulator	145
Total	1.170

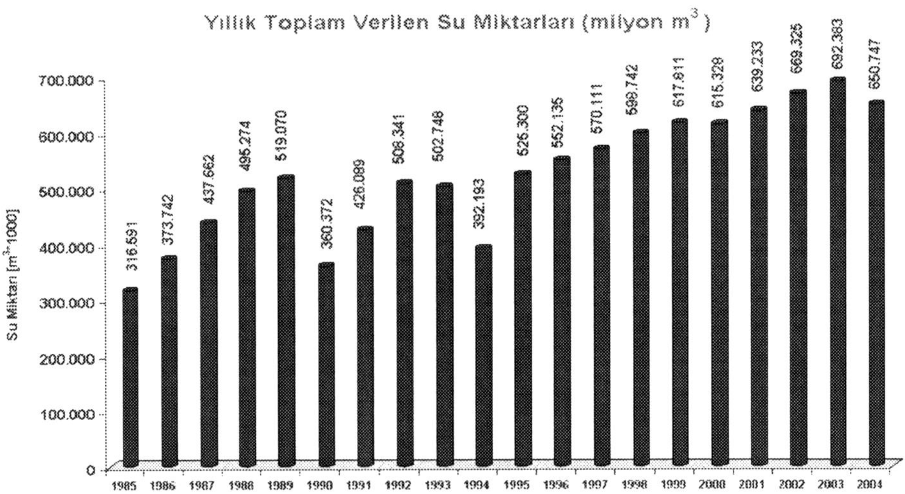

Figure 1. Annual total water distribution in Istanbul

TABLE 5. Existing water treatment plants in Istanbul

Name of plant	Service year	Capacity (m^3/year)
Omerli Orhaniye	1972	300
Kagıthane Celebi Mehmet Han	1972	378
Buyukcekmece	1989	400
Elmalı	1994	50
Ömerli Orhaniye (increase in capacity)	1995	200
Ömerli Muradiye	1995	320
Kagıthane Yıldırım Beyazıt Han (increase in capacity)	1996	70
Kagıthane Yıldırım Beyazıt Han (renewal)	1996	280
Ömerli Osmaniye (renewal)	1997	220
Ikitelli Fatih Sultan Mehmet Han	1998	420
Ömerli Emirli-Yavuz Sultan Selim	2001	500
Ikitelli 2. Beyazıt Han	2002	420
Total		3.558

TABLE 6. The amount of water treated by water treatment plants in Istanbul during 2004

Name of plant	Daily (m^3)	Annual (m^3)
Omerli	819.578	299.145.985
Kagithane	524.187	191.328.493
Buyukcekmece	229.204	83.659.700
Ikitelli	340.398	124.245.410
Elmali	20.619	7.526.224
Other	8.619	3.145.939
Total	1.942.607	709.321.103

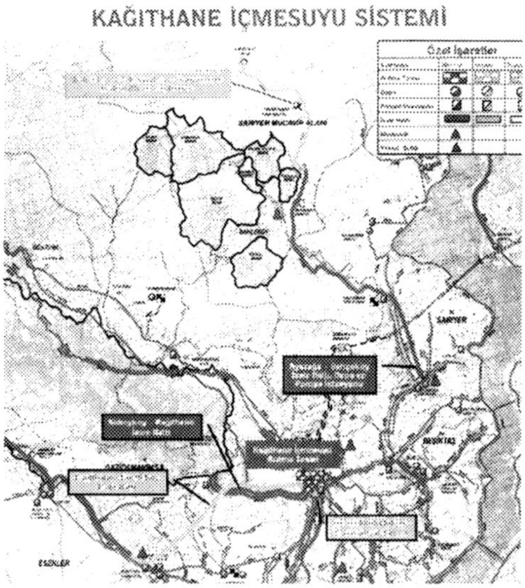

Figure 2. Kağithane water treatment system

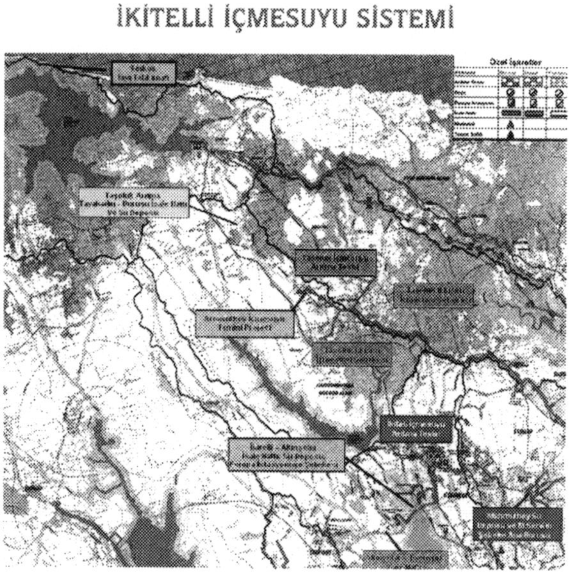

Figure 3. İkitelli water treatment system risk associated with the water distribution network of Istanbul

4. Risks Associated with Water Distribution Network of Istanbul

4.1. EARTHQUAKE

The main risk regarding the water distribution network is occurrence of an earthquake in Istanbul. During the next 25 years, the probability of an earthquake higher than 6 on the Richter scale is predicted to be 65–70%. Therefore, determination of the points of vulnerability in the water distribution pipeline system becomes extremely significant. Risks associated with the water distribution network of Istanbul after an earthquake were evaluated comprehensively in Earthquake Master Plan of Istanbul. In Earthquake Master Plan, between 1,400 and 1,600 points in water distribution network are predicted to be damaged, mostly on the European side, particularly in Fatih and Gungören districts. Besides, between 1,200 and 1,300 points in wastewater pipeline network are predicted to be damaged as well. In another study, 1,191 points are predicted to be damaged in water distribution lines, with a maximum damage at 123 points in Kucukcekmece district (Ugurlu, 2002 – from Earthquake Master Plan of Istanbul). Furthermore, 1,959 points in the wastewater pipelines are also predicted to be damaged.

4.2. OTHER RISKS ASSOCIATED WITH WATER DISTRIBUTION NETWORK OF ISTANBUL

In addition to earthquakes, there are also other risks associated with the water distribution network of Istanbul. These potential risks are: terrorist attacks to the water resources or to water distribution systems; floods and landslides; and droughts. In the metropolitan city of Istanbul, it seems clear that the whole distribution network is quite complex and exhibits a challenge for management under emergency situations.

4.3. MAIN TASKS OF AKOM AND ISKI DURING EMERGENCY CONDITIONS

AKOM (Disaster Coordination Center) was found in 2000, in order to provide support and service to the Istanbul Metropolitan Municipality in case of emergency conditions in Istanbul (www.iski.gov.tr). The main tasks of AKOM and ISKI during emergency situations are: to predict the possible damages; take precautions; and repair the damage after the disaster, to determine and provide the alternative sources of water (e.g. to design and construct water silos in the city to meet the water demand, and to make arrangements with private companies to distribute water during emergency situations, to design systems to indicate failures or damages in water and wastewater pipeline networks, and to train and coordinate the staff to be in charge during emergency conditions.

References

1. Earthquake Master Plan of Istanbul, 2003.
2. Istanbul Water Sewage Administration (ISKI) Activity Report, 2003.
3. Istanbul Water Sewage Administration (ISKI) Activity Report, 2004.
4. www.die.gov.tr
5. www.iski.gov.tr

2.4 THE WATER RESOURCES IN ROMANIA: PERSPECTIVES FOR THE SUSTAINABLE DEVELOPMENT IN ORDER TO SUPPLY WATER TO CITIES IN EMERGENCY SITUATION

LIVIU–DANIEL GALATCHI
Ovidius University, Department of Ecology and Environmental Protection, Mamaia Boulevard 124, RO 900527 Constanta - 3, Romania; e-mail: galatchi@univ-ovidius.ro

Abstract: The struggle to achieve the millennium development goals for water and sanitation in Romania will have to be achieved in cities, towns, and villages. This is where much of the industrial production and economic activities are concentrated and where most critical governance decisions are made. With the strong physical and financial growth of cities in comparison with more dispersed rural settlements, water challenges are becoming increasingly urban in nature. City and municipal governments play critical roles in water management ensuring provision of water, sanitation, and wastewater removal. How tasks related to water governance fit within broader frameworks of environmental and economic policy are of crucial importance. It is at this level that policy initiatives become reality and need both political and administrative support as conflicts have to be resolved and consensus found among competing interests and parties. Actions must be coordinated and managed in these areas.

Keywords: Romanian waters, Danube basin, water management, institutions, supply infrastructure, water collection and transfer, information, privatization, sustainable development

1. Introduction

The Danube is a 2,860 km long river that flows across nine countries (Fig. 1); its watershed of 817,000 km^2 extends on the territory of 17 countries – more than any other river basin in the world. The discharge varies along the middle and lower basin between 615 m^3/s (low water) and 8,600 m^3/s (flood, 1965) with an average of 2,400 m^3/s. The main tributaries of the Danube are (watershed in thousand square kilometers): Tisza 157, Sava 100, Siret 45, Dráva 40, Morava 38, Prut 29, Morva 27, Olt 22, Inn 26, Vah 19, and Rába 18.

There are three clearly distinguishable sections of the Danube River: (1) upper Danube, from the Black Forest to the Carpathian Mountains, where a series of hydropower stations were built slowing the flow and creating problems of sedimentation. (2) In the Carpathian basin the Danube flows through the West-Hungarian plain, meandering and, after the Danube Bend, arriving at the Great Hungarian Plain. Precious aquifers of drinking water are found here. Regulation and flood protection have an essential role and the stable riverbed is fixed by regulation works, which need constant maintenance and surveillance. (3) Lower Danube extends from the Southern Carpathians to the Black Sea. The Danube flows into the Black Sea on Romanian territory forming the Danube Delta, the largest wetland in Europe, declared a biosphere reserve of global importance due to the rich diversity of flora and fauna.

The basin contains many of the most important cities in the EU region, such as Vienna, Budapest, and Belgrade, picturesque towns like Salzburg, as well as heavy industry. Mining activities are concentrated in the mountainous regions. Several power plants (e.g. nuclear stations at Paks, in Hungary, and Cernavoda, in Romania) have been built along the Danube and its tributaries, using the river as cooling water.

Romania accounts for around 20% of the total basin area. It is located almost entirely within the basin of the Danube River, which is therefore the most important natural feature of the country.

Y. Sharan et al. (eds.), Water Supply in Emergency Situations, 49–56.

Figure 1. The Danube River basin

Romania is located in the eastern side of Central Europe. There are three important elements that define the geographical position of Romania in Europe: the Danube River, the Black Sea, and the Carpathian Mountains. Romania covers an area of 237,391 km^2. Out of the total boundary length of 3,190.3 km, the Danube River boundary represents 1,865.7 km, the territorial boundary 1,037.7 km, and the sea boundary 287.9 km. The country is drained by a hydrographical network with permanent flow of about 76,000 km, with the total length belonging to the Danube basin. Mean annual precipitation decreases in intensity from west to east, from 600 to 500 mm in the Romanian plain and under 400 mm in Dobrogea, to 1,000–1,400 mm in the mountain areas (Galatchi and Tudor, 2006).

From the administrative point of view, Romania is divided into 41 counties, plus the capital, the municipality of Bucharest. The average area of a county is about 4,600 km^2, with an average population of 500,000 inhabitants.

According to the 1999 census, Romania had a population of 22.5 million, of whom 55% are living in urban areas. At the end of 1999, the gross domestic product of the country was about $34 billion, representing a gross domestic product (GDP) per capita of $1,520. This is one of the lowest figures amongst the East and Central European countries applying for membership of the European Union (EU).

2. Water Management in Romania

Water resources management challenges differ enormously in Romania, depending on the type of human settlement. The spectrum of settlement types stretches from the very low-density-scattered single dwellings found in rural areas, through villages and small towns, to the much more dense and crowded cities. Half of the Romania's population and most of its economic output is located in urban areas. Today, large cities present a particular challenge, with many cities housing over 300,000 inhabitants.

Water resources management will always face the challenge of balancing the needs of different water users. This is the case both in large urban or relatively small rural communities. The water needs of the agricultural production, energy and industrial sectors are often in competition. So, while the overriding need to ensure adequate clean water for drinking, hygiene, and sanitation and wastewater disposal, are of paramount importance, they nonetheless need to be balanced with consideration of these and other needs.

Romania's water resources are relatively poor and unequally distributed in time and space, being formed of surface waters – inland rivers, lakes, and reservoirs, and the Danube River – and of ground waters. Black Sea water resources, although very important, cannot be taken into account for the time being because of the technical and economic difficulties in seawater desalination. The ground waters, generally of better quality than the surface waters, are estimated at an available annual amount of 9 billion cubic meters, of which about 3 billion cubic meter can be used under existing technical and economic conditions. Romania receives 85 billion cubic meter per year from the Danube River, but the possibilities for their actual use are limited because of the river's navigable character. Thus, only 30 billion cubic meter per year can contribute to the water stock that is technically available for consumption. The Danube flows alongside Romanian territory, with 37% of its length forming the southern boundary of the country. The river has already acquired important pollutants before it even reaches Romania, its water being included in pollution class II as defined by the Romanian Standard STAS 4706/1988 as it enters the country. The quality of the Danube's water is adversely affected by diffuse and point source pollution throughout its catchment area. In particular, agricultural pollution and untreated discharge from municipal and industrial sources in Romania have a negative effect, both directly from riparian sources and indirectly via its tributaries.

2.1. INSTITUTIONS

At the national level, the Romanian institutions responsible for the policy and strategy of the water sector are:

- The Ministry of Waters and Environmental Protection – responsible for the adoption of the EU water quality Acquis in Romania. The Ministry is responsible for drawing up national water policy and the preparation of legislation and regulation within this field.
- The National Company "Romanian Waters" (Apele Romane) – responsible for the enforcement of the water management policy, under the coordination of the Ministry of Waters and Environmental Protection.
- The Ministry of Public Works, Transports and Housing – responsible for the quality of construction of drinking water and wastewater treatment facilities.
- Ministry of Health and Family – responsible for drinking water quality.
- Ministry of Agriculture, Food and Forests – responsible for the use and protection of water in the agricultural field. At the local level, the Local or County Council is the authority granting development consent and is the owner or supervisor of water abstraction, supply, and treatment infrastructure.

The related environmental legislation consists of Law 137/1995 on Environmental Protection Law (framework law), Law 107/1996 on Water Law, Law 171/1997 for approval of the Plan of National Planning of the Romanian Territory – section II – Water, Governmental Decision 730/1997 for approval the NTPA 001 concerning the charging with pollutants of the wastewater discharges into water resources, Law 86/2000 for ratification of Aarhus Convention on public access to environmental information, Law 14/1995 for the ratification of Danube Convention on cooperation regarding protection and sustainable use of the river water.

Each year, the Ministry of Waters and Environmental Protection revises a Plan for Adoption of the EU Acquis in the field of environmental protection. The National Plan for Approximation of the National Legislation with the EU Acquis (finalized in May 2001) gives estimates for the transposition of the most relevant EU Directives. Table 1 shows the most expensive directives (regarding the implementation of their provisions).

TABLE 1. European Union directives (Compania Nationala "Apele Romane", 2006)

Directive	Transposition	Implementation
Drinking Water (98/83/EEC)	2000	2015
Urban Wastewater (91/272/EEC)	2000	2030
Surface Water for Drinking Water (74/440/EEC)	2000	2015
Hazardous pollutants in the ground water (80/68/CEE)	2001	2015
Surface Water for Drinking Water (74/440/EEC)	2000	2015
Hazardous pollutants in the ground water (80/68/CEE)	2001	2015
Hazardous pollutants in the surface water (76/464/CEE)	2000	2030
Public access to the environmental information (90/313/CEE)	2000	2002
Directive regarding the environmental impact (85/337/CE)	2000	2002

2.2. LOCAL WATER SUPPLY SEWERAGE AND WASTE WATER TREATMENT INFRASTRUCTURE

There is a vast differential between urban and rural settlements when it comes to centralized drinking water systems; all the 263 municipalities and towns have such systems, while only 17% of rural communities benefit from this service. The random distribution of water resources in the country's territory, an insufficient degree of regulation of river flows, and significant pollution of some inner rivers are all causes of the lack of sufficient water supply sources for parts of the country, especially during drought or very low temperature winters, when water supply can be interrupted for days and flows drastically reduced.

There is a vast differential between urban and rural settlements when it comes to centralized drinking water systems; all the 263 municipalities and towns have such systems, while only 17% of rural communities benefit from this service. The random distribution of water resources in the country's territory, an insufficient degree of regulation of river flows, and significant pollution of some inner rivers are all causes of the lack of sufficient water supply sources for parts of the country, especially during drought or very low temperature winters, when water can be interrupted for days and flows drastically reduced (Galatchi, 2006a):

- Population that benefits from both endowments – 51% of the total population.
- Population that benefits only from water supply (without sewerage) – 14% of the total population.
- Population that benefits from neither water supply nor sewerage – 35% of the total population; investments are needed to permit the gradual achievement of the standards required by the EU directives.

These will be mostly in charge of the public sector (particularly local authorities) and will be a heavy burden for the public finances. In Romania, there are now 556 operators of public services, subordinated to the local public administration authorities or with private capital, from which 74 are independent administrative structures and 482 are commercial companies. The private capital involvement and the achievement of strong and lasting partnerships between public and private sector are now at the beginning. There have already been some examples of privatization in this field, such as in Bucharest and Ploiesti, for the water supply and waste water system.

Main problems and issues are:

- Lack of information, awareness, and interest concerning water issues both at the general public and decision makers level.
- Privatization.
- Inadequate transboundary cooperation between neighboring countries (especially, Romania, Hungary, Bulgaria, and Yugoslavia).

2.3. WATER COLLECTION AND TRANSFER IN ROMANIA

The evolution and development of the water supply system is closely connected to the sewer system in Romania (Table 2).

TABLE 2. Water supply and sewage systems in Romania (Rojanschi, 1999)

Type of human settlements	Number of human settlements	With water supply systems		With sewage systems	
		Number	Percentage	Number	Percentage
Large cities	262	262	100.0	261	99.6
Rural community centers	2,686	1,249	46.5	330	12.3
Villages	10,390	1,045	10.0	16	0.2

Currently, 261 large cities (without Fundulea City) and 346 villages (2.6% out of the total number) are connected to sewer networks. In Braila and Harghita counties, no village has a sewage network, and Giurgiu and Ialomita counties have only one a piece each. Around 47 of the largest cities which do not have wastewater treatment plants (Bucharest, Braila, Craiova, Turnu Severin Tulcea, etc.) produce ~20 m^3/s of wastewater, which is directly discharged untreated into surface water. In the localities without centralized sewage systems, the wastewater will either be collected in waterproof basins with periodical emptying and transfer into wastewater treatment plants, or it will be discharged into individual absorbing septic tanks.

In Romania, most sewer pipes or channels are made of bricks (the older ones), reinforced concrete, or different plastic materials (more recent ones).

The amount of wastewater treated to meet national norms is unsatisfactory, so the sewer systems are pollution sources for the surface and ground water. The most frequent contaminants are insoluble inorganic salts and oils.

The actions envisaged for the rational management of waters in Romania will result in securing good-quality water resources for the population and the necessary resources for the economic activity, especially for industry, agriculture, an update of the system of alarming and warning the population about potential hazards, the starting of a process of investment in protection works against floods and dangerous meteorological phenomena.

The main measures in the government program are aimed at:

- The rational management, in terms of quantity and quality, of the surface and ground waters, in view of ensuring the water resources for various uses, through hydrotechnological works that should have only a minimal negative impact on the environment.
- The improvement of the quality of river waters, following the operation of the city wastewater treatment plants in Bucharest (Glina), Braila, Galati, and Tulcea.
- Works on sewerage systems and wastewater treatment plants in urban and rural localities, according to the program of activities.
- The completion of works on the water supply facilities intended for urban and rural localities, including the rehabilitation of the water treatment plants for the cities of Bucharest, Timisoara, Iasi, Constanta, Ploiesti, Brasov, Cluj, Oradea, Baia Mare, and Bacau.
- Continuation of the adjustment, storage, and development works in the hydrographical basins (the completion of the complex water supply systems of Zetea–Dumbraveni–Medias–Copsa Mica, Prahova Valley–Azuga–Busteni–Sinaia–Comarnic–Breaza, the depollution, cleaning, and adjustment of the Sasar River in Baia Mare).

- The rehabilitation and protection of the Romanian Black Sea Coast and the offshore area belonging to Romania, considering the priorities of the National Strategic Plan for the rehabilitation and protection of the Black Sea.
- The remaking and conservation of the seaside area and of the lakes with therapeutic properties (Techirghiol, Amara, etc.), the protection of the Romanian Black Sea Coast against erosion.
- Protection against floods and dangerous weather phenomena, through planning works on the river course (including the rehabilitation of the dams on the Danube), and through an upgrade of the information system of warning and alarming the population.
- The harmonization of water-related legislation with the provisions of the new Directive of the European Parliament and Council of Europe concerning the community action framework on water policies (Guvernul Romaniei, 1999).

2.4. INFORMATION AND AWARENESS IN WATER ISSUES AMONG THE PUBLIC AND DECISION MAKERS

The general public is not aware of the fact that water is not an inexhaustible resource. They also do not have enough information about the consequences of water pollution and waste. Water treatment requires large-scale financing and waste results in higher prices for the consumer, which, in the absence of information, can cause consumer anger and mistrust between the people and the authorities. The price of water is still very low (the lowest in the total household expenses, compared to the electricity or gas prices), therefore, people do not pay enough attention to consumption rate, and a lot of water is wasted.

In the framework of a European Bank for Reconstruction and Development (EBRD) loan in the Municipal Utilities Development Programme (MUDP) I and II, several big cities in Romania (Iasi, Timisoara, Craiova, Targu Mures, Brasov) benefit from investments in water infrastructure rehabilitation and modernization. Water meters have been installed and the price of water increased, as a condition of the Loan Agreement between EBRD and the Government of Romania. Since there was no proper public information program, this new situation generated anger and some conflicts occurred between the local population and water services operators. The results were unpaid water bills and even disconnection from the water distribution network.

The right of the person to have access to any public interest information can only be limited in special circumstances. In this respect, public authorities, according to their respective competencies, are required to ensure that citizens are informed about matters of public and personal interest. Romania has ratified the Aarhus Convention on public access to environmental information. Further to Aarhus, the right to environmental information is translated into a number of laws and regulations that specify the responsibilities and requirements for information provision. The water and environmental protection laws of Romania make provisions for public participation. The law requires public consultation of water users, riverside residents and the general public on all matters that affect their interests. Any decision should be taken only after having consulted these persons. However, despite these provisions, little consultation is actually carried out by the authorities.

A particularly important necessity is the provision of useful, up to date and filtered information for local government officials, who are not able to use either EU or national funds and training facilities for these issues.

Problems to be addressed include (Galatchi, 2006b):

- The need for a guide presenting the main steps to reach available funding sources for water infrastructure investments; in small and medium towns – this information is often unavailable.

- Lack of awareness among many local decision makers who are not advised of the present status of water legislation.
- Difficulties and expenses faced in achieving the high water and environmental standards now sought in the region.
- Lack of communications between local and regional authorities and their counterparts in other parts of the country and basin which denies them the opportunity to gain from others' experiences and the ability to avoid duplicating mistakes; it also becomes dangerous for the environment and public health at times of transboundary emergency, and heightens the risk of conflicts – cooperation and communication could greatly reduce the expense of inefficient water management.

2.5. PRIVATIZATION

Informing citizens of their rights and encouraging them to participate in the decision-making process by enhancing dialogue with their local politicians will encourage the individuals in local communities to use their power as voters and consumers in order to ensure that privatizing companies fulfill their contractual obligations. At the same time, providing useful and objective information to local and regional authorities will enable them to carry out their role as regulators more effectively. Public authorities are often no match for major international private water companies in the negotiation of contracts and in regulating company activities; civil society should be informed and actively questioning this issue and should insist on effective regulation by their elected officials and full accountability of the private sector.

In Romania, at the local level, the Local or County Council is the authority granting development permits and is the owner or supervisor of water abstraction, supply, and treatment infrastructure. The rehabilitation and improvement of water infrastructure is a process that will no longer be supported by the national budget; it therefore requires funding and support through private capital. Privatization of water services only began in 1999. The process, developed in Bucharest (concession contract completed in 2000) and Ploiesti, was very long and difficult, mainly because the authorities were not well informed or prepared. They did not have enough information about the national legislation in force and its compliance with EU legislation, private–public partnerships, and available funding sources for investments in water infrastructure or different privatization models and contracts.

The local authorities and their subordinate institutions need to have a clear idea of the risks and benefits, both political and economic, of investing in the water sector, either by inviting companies to participate in the privatization or by using their own means to improve water services.

Water management requires close cooperation and information exchange between the countries of the Danube River basin. The decision makers have to take into account the possibilities presented by EU adhesion and the water management directives of the EU, which highlight transboundary cooperation of whole watersheds. It would be useful to look at the example of transboundary pollution management in the Rhine basin, which was belatedly put into practice also after a major industrial accident.

3. Concluding Remarks

There have been significant trends in Romania towards increased urbanization, as people are widely migrating from rural to urban settlements. Not so heavily reported, but nonetheless significant, is the growth of large numbers of smaller cities and towns, most of which are feeling the strains of rapid expansion. In the developing world, with total populations set to increase, overall rural populations are expected to remain largely unchanged in numbers, while urban populations are expected to grow rapidly. In some of the more developed countries where the vast majority of the population lives in cities, there are signs of counter migration: people abandoning the cities for better living standards in surrounding smaller communities.

As human settlements are the major polluters of water resources, good water and wastewater management is essential to limit pollution and minimize health risks. The expansion of urban areas and agricultural frontiers usually present new opportunities for disease. Furthermore, as industries tend to be concentrated in or around cities and agricultural production predominantly in the surrounding available areas, measures to stem pollution and introduce and maintain efficient and safe drinking water, as well as wastewater disposal mechanisms must be extended. This is essential to ensure the health of populations and particularly the inhabitants of large urban communities. Failure to meet these challenges will have a disastrous effect on the further expansion of cities.

Acknowledgment

We thank Professor Dr. Stoica-Preda Godeanu for helpful suggestions during our research.

References

1. Compania Nationala Apele Romane (2006) Internal report of the "Romanian Waters" National Company, Bucharest.
2. Galatchi, L.D. (2006a) The Romanian national accidental and intentional polluted water management system, in: Dura, G. et al. (eds) Management of Intentional and Accidental Water Pollution, Springer, New York, pp. 181–1184.
3. Galatchi, L.D. (2006b) A Long-term perspective on the sustainable development of urban groundwater resources in Romania, in: Tellam, J.H. et al. (eds) Urban Groundwater Management and Sustainability, Springer, New York, pp. 129–136.
4. Galatchi, L.D. and Tudor, M. (2006) Europe as a Source of pollution – the main factor for the eutrophication on the Danube Delta and Black Sea, in: Simeonov, L. and Chirila, E. (eds) Chemicals as Intentional and Accidental Global Environmental Threats, Springer, New York, pp. 57–63.
5. Guvernul Romaniei (1999) Strategia Nationala pentru Dezvoltare Durabila, Bucharest, pp. 29–32.
6. Rojanschi, V. and Diaconu, G. (1999) Ingineria mediului, Universitatea Ecologica, Bucharest, pp. 42–122.

CHAPTER III

SECURITY OF WATER SUPPLY

3.1 SECURITY OF WATER SUPPLY AS A CATALYST FOR CONFLICT RESOLUTION IN THE MIDDLE EAST

ASHER VATURI [1] AND ABRAHAM TAL [2]
The Interdisciplinary Center for Technological Analysis and Forecasting, Tel Aviv University, Tel Aviv, Israel

Abstract: Water resources are in many countries transboundary and as such, especially in arid regions such as the Middle East, may provide a basis for conflict. War almost broke out between Syria and Iraq following the construction of a dam on the Euphrates River by Syria. This is even more critical in cases where conflict exists due to other reasons. Such is the case with the Israel and the Arab neighboring countries conflict with the water issues between Israel and the Palestinian Authority being particularly complicated since they are intertwined in all aspects.

The danger of water issues becoming *casus belli* has triggered water allocation and/or treatment arrangements between Israel and each of its neighbors. One of the more important issues in securing the Israel and the Kingdom of Jordan peace treaty was the sharing of water resources, following many years of trust-building agreements even though the two countries were in a state of war. Between Israel and the Palestinians, cooperation in water issues is essential for both sides and thus, even through the difficult periods of *intifada* (uprising) the only cooperation which existed was in the water sector. Several joint projects have and are being carried out and joint committees meet to solve problems.

Transboundary-shared water resources may be a reason for conflict; however, if handled wisely by the involved countries, may even provide a basis for settling other issues of conflict.

Keywords: water resources; Middle East, transboundary, conflict, conflict resolution, cooperation

1. The Middle East Water Issues

Water resources are worldwide regarded as a basic component of human security. It is more significant in the region of the Middle East (Fig. 1) where the water resources have always been limited (Fig. 2). Moreover, growing populations place ever-increasing demands on the limited resources and cause increasing scarcity of water. Such circumstances have traditionally caused regional and international conflicts.

This condition applies to the Middle East ecosystems – arid areas which populate some 6% of the world population, but have only some 1% of its renewable fresh water. Shortage of water is an acute issue not only in agricultural countries such as Egypt, Jordan, and the Palestinian Authority, but also in more industrial country such as Israel.

Topkaya [1] argues that as unused water resources in a region decrease, water for one user means lack of water for the other. Such situations in the Middle East, in recent years, have led

[1] Asher Vaturi, Researcher on Environmental issues, ICTAF at Tel Aviv University, Tel Aviv 69978, Israel, e-mail: asherv@eng.tau.ac.il
[2] Abraham Tal, Head Space, Environment and Remote Sensing Division, ICTAF at Tel Aviv University, Tel Aviv 69978, Israel, e-mail: atal@eng.tau.ac.il

Y. Sharan et al. (eds.), Water Supply in Emergency Situations, 59–64.

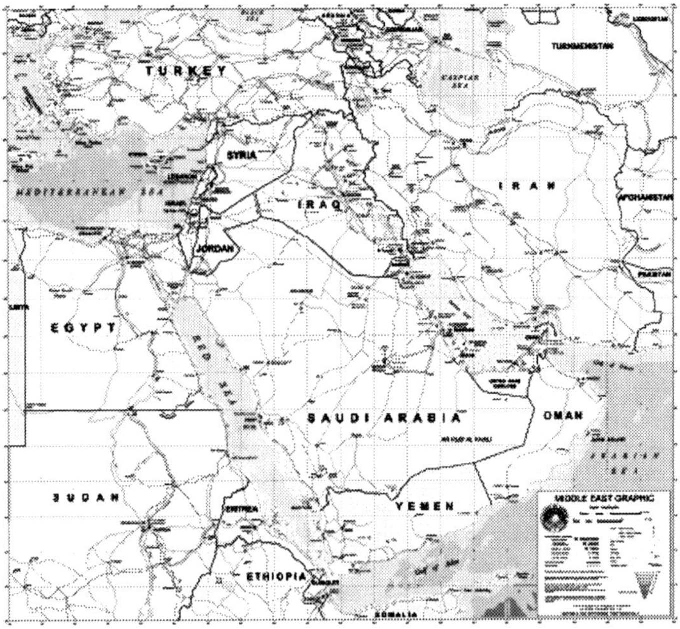

Figure 1. The Middle East

to competition over the available resources and in certain cases to mobilization of armed forces and to conflicts between neighboring countries.

Theoretically, the allocation of water resources was arranged by the international law. In 1910, the Institute of International Law proposed a framework for regulating international waterways. In the 1920s, the League of Nations adopted the only two existing multilateral treaties on the use of international waterways [2].

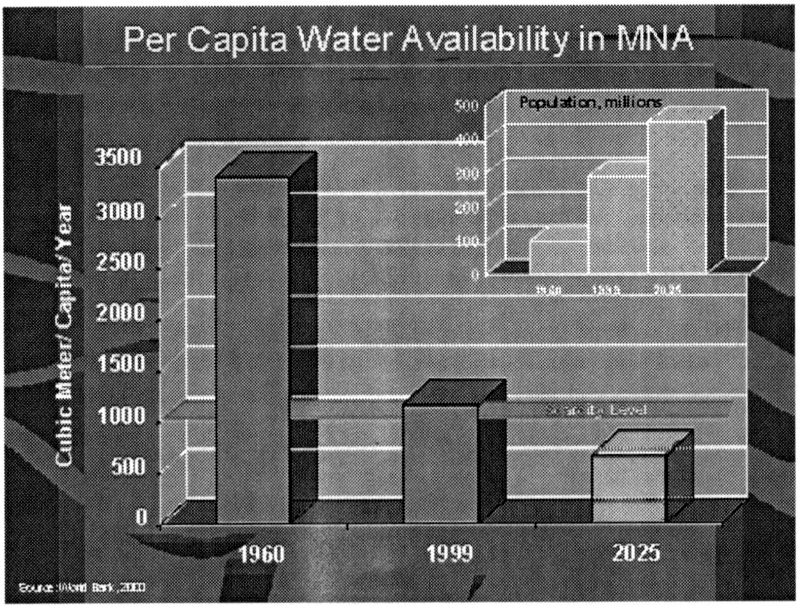

Figure 2. Per capita water availability in the Middle East and North Africa (MNA). (From World Bank.)

The main progress regarding the principles of international law relating to transboundary water resources was established in 1966 in Helsinky by the International Law Association's (ILA) Helsinki Rules on the Uses of the Waters of International Rivers [3]. Further progress of the international legal framwork of crossboundary water resources was made during the "UN Convention on the Law of the Non-Navigational Uses of International Watercourses" in 1997.

However, the international legal framwork is still incomplete and the uncertainties have lead to local and nongovernmental initiatives aimed at finding peaceful solutions for water allocation, securing water supply to all parties.

Unlike military security, human security is a very broad issue, based on justice, personal safety on basic human rights. However, not less important are such notions as compassion, common understanding, and trust. Regarding water resources as a prime example, this issue requires commitment to solutions that transcend political boundaries.

Policy makers have in many countries found themselves under intense political pressures to take measures to secure national water resources. However, these measures often have an impact on the water supply of another country, which leads to difficulties in international relations.

This has in many cases led to situations where despite its traditional role as a cause of conflict, shared water resource have created possibilities for cooperation.

The immediate conclusion is that a regional water management system is essential for both secure of supply and for eliminating unnecessary conflicts. Moreover, successful water cooperation has a potential for further regional cooperation in other areas.

1.1. CONFLICT BETWEEN SYRIA AND IRAQ

Indeed, there are examples for military conflicts as a result of competition over water resources in the Middle East. For example, Syria's construction of the Al-Thawra dam on the Euphrates River in 1974 created a political crisis with Iraq, because as a result, the river's flow was reduced by a quarter. Iraq threatened to bomb the dam and massed troops along the border. The crisis was resolved by mediation. Also the 1967 war between Israel and Syria followed Syrian attempts to divert some of the Jordan River sources north of the Israeli boundaries.

1.2. WATER COOPERATION AS A TRIGGER FOR PEACE

Nevertheless, water supply issues in the Israeli–Arab conflict have been primarily the focus of cooperation rather than conflict. There are many examples that support this point. Much of the common understanding obtained in the Arab–Israeli conflict on water supply issues is based on the nature of the water resource being understood by all sides as essential to life.

The peace treaty between Israel and Jordan is a good example. Even for several decades prior to the signing of this treaty, water negotiators from the respective sides were meeting and discussing the allocation of water rights relating to the two shared rivers, the Jordan and the Yarmouk. The agreements and the arrangements on these issues seem to have been the first steps of trust building between the two countries, which lead to the relationship needed to make peace happen years later.[1]

2. Water Issues and Cooperation between Israel and the Palestinian Authority

The water issue was one of the main topics in the peace talks between Israel and the Palestinians. It is indeed worthy of mention that the Israeli–Palestinian Water Committee is the only working group that survived the collapse of the Oslo Peace Accords – continuing to meet to this day. Moreover, the only agreements signed by Israel and the Palestinian Authority during the second Intifada have both involved water issues – an agreement to refrain from causing damage to each others water infrastructure and an agreement on standards for sewage treatment.[2]

2.1. THE PERES CENTER FOR PEACE PROGRAM

There are in general two main strategies to achieve this kind of regional water cooperation. The first involves the use of advanced technologies to develop water sources and the second relates to regional programs for efficient use of water. These strategies were adopted by The Peres Center for Peace in order to promote a long-term solution of the conflict between Israel and its neighboring countries, focusing mainly on large-scale water treatment and desalination [4].

The program of Peres Center is based on the idea that any initiative should take into consideration the potential profits of all parties involved. It argues that water saving by efficient use might be defined as a new source of water for the welfare of all parties. This additional water can be gained by reduction of water losses through conveyance system restoration. Furthermore, high-tech agricultural technologies developed in Israel are and can be put to use in rural areas, mainly in Jordan and the Palestinian Authority. The said technologies enable a high agricultural rate of production per cubic meter of water and can be accomplished with the existing agricultural infrastructure.

The initiators at the Peres Center believe that such a technology transfer could change the current situation in most of the Middle East countries where agricultural development is limited by shortage of water. The strategy of the center is to accomplish increased water efficiency by substitution of the physical amount of water with advanced agricultural procedures including irrigation management and technology, crop selection, plant genetics and development of new varieties, protected soil-less greenhouse agriculture, planning, and training.

2.2. THE "GOOD WATER NEIGHBORS" PROJECT

Another project using water shortage as a catalyst for dialogue, is the "Good Water Neighbors" (GWN) project (Fig. 3), which was established by EcoPeace/Friends of the Earth Middle East

Figure 3. Good Water Neighbors

(FoEME) in 2001 [5]. The aim of the project is to promote understanding and to raise awareness of mutual water problems across the borders by Palestinians, Jordanians, and Israelis.

The methodology used in this project is based on an original idea of identifying cross-border communities and utilizing their mutual dependence on shared water resources as a basis for developing dialogue and cooperation on sustainable water management. The organizers have succeeded to create an improvement within the water sector by building trust and understanding that has led to common problem solving and peace building among communities even in the midst of conflict.[3]

The project developed not only on a background of a long tradition of conflict, but also of cooperation between Israel and the Arab countries in the area of water supply. It is based on fairness, justice, and on a reasonable allocation and sustainable management of the water resources.

Other important problems in the water sector between Israel and its neighbors include the lack of sewage treatment, overpumping of aquifers, excessive diversion of surface water flows, and difficulty in implementing critical water-demand management policies. These conditions threaten scarce water resources, pose environmental and health hazards to both Israeli and Arab communities, and create a potential for additional cross-border conflicts. The initiators of the project, EcoPeace and FoEME focused on the community level, and on fostering the cross-border relationships that are necessary to solve common water problems.

The first phase of the project was implemented between 2001 and 2005 and involving initially 11 selected Israeli, Palestinian, and Jordanian communities.

Today, in second phase of the project, 17 communities are involved. Each community works together with a neighboring community on the other side of the border on common water issues. On the local level, the project involves community representatives including youth groups to improve their own water quality and quantity through education and awareness activities.

In each community water-saving devices were installed in all public buildings and schools were transformed into water-saving model buildings. The water trustees themselves carried out surveys of all water taps in public buildings within their community and then, by installing the devices, cut by one third of the amount of water used in public buildings.

2.3. TRANSBOUNDARY WATERSHED ANALYSIS

The Interdisciplinary Center for Technological Analysis and Forecasting (ICTAF) has a long record of cooperation with Palestinian and other Arab country researchers in different EU- and US-supported projects. In this context, worthy of mention is a joint ICTAF/ARIJ[4] study analyzing all the water-related issues in a joint, cross-boundary watershed – that of Wadi Zimmer/ Nachal Alexander which starts out at Nablus and emits into the Mediterranean in central Israel. The study assembled and analyzed all available data sets on water sources, usage, sewage, agriculture, industry, settlements and so on. This study was funded by USAID [6].

3. Conclusions

Water, being essential to life and when in shortage, may be the cause of conflict and even of *casus belli*. Since in most of the countries of the world, at least part if not all, the water resources are transboundary, cooperation in using and conserving the resources is an essentiality for good and peaceful relations. This is even more pronounced where there are other reasons for conflict. Therefore, basic water policies at the national level, must address the need and ways of cooperation between the partners to the shared water resources.

References

1. Topkaya B. (1998) Water resources in the ME: forthcoming problems and solutions for sustainable development in the region, Akdeniz University Faculty of Engineering, Department of Environmental Engineering, Antalya, Turkey.
2. Lazerwitz, D.J. (1995) The flow of international water law: the international law commission's law of the nonnavigational uses of international watercourses. Global Legal Studies Journal.
3. ILA (International Law Agency) (1967) Helsinki rules on the uses of international waters of international rivers, Report of 52nd Conference, Helsinki.
4. Peres center (2006) Regional agriculture, water and food security programmes, The Andreas Agricultural Development Trust.
5. Friends of the Earth (Israel) (2006) http://www.foeme.org/projects.php?ind=54
6. Tal, A., Albert J., and Rishmawi, K.N. (2001) Development of a Spatially Referenced Database for Management of an Israeli–Palestinian Transboundary Watershed, ICTAF/ARIJ Report to the US Embassy Office of the Science Counselor, August 2001.

3.2 CHERNOBYL EXPERIENCE FOR WATER PROTECTION IN CASE OF RADIATION-RELATED TERRORIST ATTACKS

ZLOBENKO BORYS*
Institute of Environmental Geochemistry of NAS & ME of Ukraine, 34A, Palladina Ave., Kiev 03142, Ukraine

Abstract: From the first day following the accident at Chernobyl Nuclear Power Plant (ChNPP), one of the most serious problems was to prevent general contamination of the Dnieper water system and to guarantee safe water consumption for people living in the affected zone. The water protection and development of monitoring programs for the affected water bodies were among the most important postaccident countermeasures taken by the Government Bodies in Ukraine. To solve the water quality problem for Kiev, an emergency water intake at the Desna River was constructed within a very short period. Since the accident, 192 groundwater wells (out of 368 constructed) have been supplying potable water to the city of Kiev through a series of pumps and reservoirs. The most effective measures to reduce radioactivity in drinking water are those, which operate at the water treatment and distribution stage.

Keywords: Chernobyl, urban water supply system, terrorist acts

1. Introduction

The real threat of nuclear, biological and chemical weapons did not disappear with the termination of the Cold War. The accident at Chernobyl Nuclear Power Plant (ChNPP), terrorist acts in big cities of the Europe and the USA, and the threat of a "dirty bomb" demonstrated to the whole world, that large territories, including big towns with multimillion populations could be affected by intensive radioactive contamination. Many factors may disrupt regular water supply: destruction of the infrastructure following a natural disaster, contamination through accident, or an act of terror.

During the initial accidental release, the surface water bodies around the Chernobyl NPP were directly contaminated by atmospheric fallout [1]. The former Soviet Government paid special attention to prevention of the large-scale contamination of the Dnieper water system by Chernobyl radionuclides and providing people living in the affected zone with clean drinking water and safe water use. The strategy of aquatic countermeasures defines many mitigation actions, which were applied on the water intakes and irrigated channels. The main objective of water remedial activities implemented since 1986 was to prevent significant secondary contamination of the surface water bodies and to mitigate expansion of expected ground water contamination.

2. State of the Art of Water Use

Ukraine is heavily dependent on surface water resources for drinking water supplies. The Dnieper River system is the main water-supply of Ukraine. More than 80% of total water consumption stems from surface water bodies (rivers, lakes, and reservoirs). Potential surface water resources are estimated at 210 km^3, whereas potential groundwater resources are estimated at 22 km^3.

* To whom correspondence should be addressed. Borys Zlobenko, Institute of Environmental Geochemistry of NAS & ME of Ukraine, 34-A, Palladina Ave., Kyiv 03142, Ukraine, e-mail: borys_zl48@yahoo.com

Y. Sharan et al. (eds.), Water Supply in Emergency Situations, 65–70.

Kiev is one the largest cities in Europe and the capital of Ukraine. It occupies the area of 408 km^2. Its population is 2.63 millions. At present surface water resources of Kiev (the Desna and Dnieper Rivers) easily meet the water supply needs of the city.

The rivers Dnieper and Desna are not protected against occasional or accidental pollution by liquid effluents from industrial enterprises, river transportation of chemical substances, etc. In periods of higher turbulence of the river flow the concentration of radionuclides in Dnieper is increased (the aftermath of the Chernobyl disaster), but do not exceed the established norms. Use of waters exposed to radioactive contamination is conducted under the Law of Ukraine [2] "On legal regime of the territory exposed to radioactive contamination as a result of Chernobyl accident", and other acts of the legislation of Ukraine (Article 84 Use of Waters Exposed to Radioactive Contamination, The Water Code of Ukraine [3]).

Considerable groundwater sources are available for water supply utilization. Within the city itself, there are also large reserves of groundwater, which are used as the source for water supply. At present there are 368 municipal wells and about 300 wells, which are used for industrial purposes. Water in these wells is abstracted from two aquifers, which lie at depths of 100–300 m under the entire territory of the city.

Kiev Vodokanal [4, 5] (KVK) has 303 production wells with a total capacity of 434,000 m^3/day. Today, only 192 wells are exploited with a total production of about 240,000 m^3/day. The raw water quality of groundwater sources is monitored continuously on a daily basis.

Treated water is pumped from three main pump stations to the water distribution network, which has a series of reservoirs. The water storage reservoirs' total capacity is 450,000 m^3 that corresponds to 7.5 h of average water consumption in Kiev. The water distribution network of KVK comprises about 3,708 km of pipes, 73% of which are cast iron pipes, the rest are steel pipes.

The present urban water supply system is characterized by insufficient water resource protection, partly due to insufficient sludge treatment and disposal. The present rural water supply system is, among others, characterized by:

- Low coverage with centralized water supply systems.
- Very high noncompliance to chemical and bacteriological parameters in shallow.
- Wells owing to pollution of groundwater and surface infiltration into the dug wells.

There is a difference in quality between water from centralized and decentralized supply systems. Water quality in decentralized systems is the worst (some 30% of samples fail to meet standards, compared to 5.7% in the centralized supply). The major reason for this is that 55.2–64% of these systems have no protected sanitary zones around their sources and 27% – no disinfecting facilities. Around 9% of the centralized water-supply systems and 29% of the decentralized supply systems do not meet microbiological standards.

In January 2002, the Parliament adopted a state program for the water sector development to be implemented in two phases (2002–2006 and 2006–2011). The program is aimed at improving drinking water quality, promoting environmental friendly usage of water resources, including optimization of water balances, and ensuring proper protection and restoration of water resources.

After the Chernobyl accident in accordance with Soviet regulations the Ministry of Health of the USSR defined in 1986, 1988, and 1991 the temporary acceptable levels (TAL) for radionuclides in foodstuff and potable water (Table 1). In 1997, the Ukrainian Ministry of Health approved acceptable levels for ^{137}Cs and ^{90}Sr in foodstuff and potable water (AL-97).

TABLE 1. Temporary acceptable levels

Name of the product	^{137}Cs	^{90}Sr
Bread and bread products	20	5
Potatoes	60	20
Vegetables (root, leafy)	40	20
Fruits	70	10
Meat and meat products	200	20
Fish and fish products	150	35
Milk and milk products	100	20
Eggs (per piece)	6	2
Water	2	2*

*4 Bq/l until 01.01.1999

This document was introduced into practice from January 1, 1998. Introduction of AL-97 provides guarantee that annual individual effective dose of internal irradiation due to consumption of food products does not exceed 1 mSv for any person.

The Law on Sanitary and Epidemiological Protection of the Population of 1994 [6] defines the rights and responsibilities of industrial facilities, agencies, organizations, and individuals, and establishes the organization and procedures of the state sanitary epidemiological service and control. The most important difference compared to the former Soviet law is that industries, agencies, and organizations are responsible for informing the state sanitary epidemiological stations about any hazardous substances or health risk.

The Sanitary epidemiological stations draw random samples from 1,139 municipal, 6,899 departmental, and 8,179 rural pipes, and from 158,254 points of decentralized water supply, including 152,440 wells, 996 springs, and 4,818 artesian wells. The sanitary regulations differentiate between those of centralized and those of decentralized sources (state sanitary norms and regulations on drinking water hygiene requirements of the quality of centralized water supply systems [7], No 383 of 23 December 1996 and No 136/1940 of 15 April 1997).

3. Current Law on Emergency Measures

In case of threat of a disaster related to adverse effect of waters, local Radas of people's deputies, with involvement of enterprises, institutions, and organizations, are bound to take emergency measures on prevention of the disaster, and in case of its coming, measures on urgent elimination of its consequences, according to the legislation of Ukraine [3].

In case of emergency situations related to contamination of water bodies, that can adversely impact people's health and the state of water ecosystems, enterprise, institution, or organization which is responsible for the accident or which discovered it should urgently start to eliminate its consequences and inform state bodies on environmental protection, sanitary supervision, water resources, and relevant Rada of people's deputies.

State bodies and relevant Radas of people's deputies undertake measures on providing stable water supply of population and industries.

The State Committee for Hydrometeorology of Ukraine carries out observations on disaster development, and provides local Radas of people's deputies and state executive bodies with necessary hydrological and hydrochemical information.

Local Radas of people's deputies are bound to inform population on emergency, disasters, their scales, possible violation of ecological safety, and about measures undertaken to eliminate their consequences.

4. Brief Analysis of Emergency Phase

From the first day following the accident at ChNPP, one of the most serious problems was to prevent general contamination of the Dnieper water system and to supply people living in the affected zone with safe water for consumption. The water protection was one of the significant directions of the post-Chernobyl accident countermeasure activities in Ukraine [8–10].

The Dnieper River crosses the Ukrainian territory from its border with the Russian Federation and the Republic of Belarus in the north to the Black Sea coast in the south. ChNPP is situated near the bank of the Pripyat River 30 km from its outflow to the Kiev Reservoir of the Dnieper River.

During the initial period of the Chernobyl nuclear accident, the surface water bodies were directly contaminated by atmospheric fallout. Regular sampling was carried out at the water bodies of Chernobyl near zone, at all reservoirs of the Dnieper cascade and also at the Belorussian part of the Pripyat River basin. The special water-sampling program was implemented in the Chernobyl exclusion zone on a regular basis to control the radionuclide dispersion from this territory via the Pripyat River and into the whole Dnieper basin [11]. Development and sustainable support of monitoring programs for the affected water bodies during the whole postaccident period was one of the most important tasks for governmental bodies.

Such monitoring actions and mathematical modelling results made it possible to obtain reliable data on contamination of the sources for Kiev water supply and for other principle water intakes from the rivers and reservoirs affected during Chernobyl accident. It also allows justifying some restrictions and recommendations aimed at eliminating radionuclide migration within aquatic pathways.

The high radiation level characterized surface water contamination over the wide spectrum of short-lived radionuclides. The total beta contamination of the open water bodies near the ChNPP reached approximately 10^{-6} Ci/L. The beta activity of the Pripyat River water downstream of ChNPP in early May 1986 exceeded 10^{-8} Ci/L. The range of radioactivity in Dnieper River water near the main water intake of Kiev city (at 130 km down flow from the ChNPP) was from 10^{-10} to 10^{-8} Ci/L in May and June 1986. The largest contribution to water contamination in first month after the accident was from ^{131}I. Since 1987, ^{137}Cs and ^{90}Sr had the largest influence on the water contamination [12].

The ground water being significantly polluted in the neighboring areas of ChNPP and at short distances from the waste disposal sites contributes no more than 2–3% of the total transfer (washout) from terrestrial environment. However, groundwater contamination in this area continues to be under long-term control as well.

5. Water Protection Activity and Countermeasures

The water protection and development of the monitoring programs for the affected water bodies were among the most important postaccident countermeasures taken by Government Bodies in Ukraine. Such monitoring actions made it possible to obtain reliable data on contamination of the sources for Kiev water supply and for other principle water intakes from the rivers and reservoirs affected during Chernobyl accident.

To solve the water quality problem for Kiev, an emergency water intake at the Desna River was constructed within a very short period. The emergency intake comprised a floating water intake station "Rosa" and a main water line from the intake to the Dnieper Water Treatment Plant. The intake and water line still provides the possibility of abstracting water from either the Dnieper or Desna Rivers. Since the accident, 192 groundwater wells (out of 368 constructed) have been supplying potable water to the city of Kiev through a series of pumps and reservoirs.

Since the accident, engineering and administrative countermeasures have been taken to mitigate risk for the population that resides along the Dnieper reservoir system downstream of Chernobyl. Countermeasures during the emergency phase were based mainly on administrative decisions and were aimed at controlling the situation. These countermeasures included:

- Restriction of water use and fishery.
- Additional purification of drinking water in urban water treatment plants.
- Increased use of groundwater sources by municipalities and construction of supplementary groundwater supply wells.
- Cleaning of wells.

The main purpose of the remedial activities was to prevent significant secondary contamination of surface water bodies that are hydraulically connected with the contaminated area and to mitigate increase of expected ground water contamination. Most of the carried out water protective countermeasures were applied to the Chernobyl exclusive zone. Many other mitigation actions were applied on the water intakes and irrigated channels.

The following water protection activities were conducted [13, 14]:

- A system of clean dams was set up to cut off transport of radionuclides from the ChNPP site.
- Clean and filtrating dams were constructed on small rivers and watercourses of the 30 km zone of the ChNPP to reduce transport of particulate radioactive material.

During 1986 and the early months of 1987, over 130 special filtration dams [9, 14] with sorbing screens containing zeolite (klinoptilolite) were installed for detaining radionuclides while letting the water through. More or less effective zeolite filtration dams captured the short-lived radionuclides during the summer 1986. Very soon after, their adsorption capacity decreased because the pores of zeolite bodies were blocked by suspended matter and on other unforeseen reason. After the spring flood of 1987, the construction of new dams was terminated and the decision was made to destroy most of the existing dams. It was found that the ^{90}Sr concentration reduction by the dams studied was insignificant (the average of Df = 1.2). The effect of reducing the ^{137}Cs concentration in particulate material was not observed. The estimation indicates that since the protection facilities were put into operation in May 1987 the zeolite of all the facilities accumulated only about 1% of the radionuclide activity transported by the Pripyat River. The major part of the activity (80%) in the flood period was found to be bound to the fine particulate material (particle size less than 0.05 mm) [15].

These countermeasures required large financial and human resources for their implementtation. Although some countermeasures and cleanup activities applied to radionuclides sources on catchments proved to have positive effects, many other actions were evaluated as ineffective and even useless. The priority and available technologies for the water remediation have changed over time.

The errors in decision making were caused by inadequacy of information on spatial and temporal variations of water bodies' contamination. The price paid by the society for nonpreparedness to solve the problems of water quality management at the polluted regions was high enough [16]. As a result, many useless water protective actions were taken during the first month after the accident.

6. Conclusion

The intervention measures to reduce contamination of rivers and reservoirs are thought to be unpractical and of low efficiency at reducing doses through the drinking water pathway.

The most effective measures to reduce radioactivity in drinking water are those, which operate at the water treatment and distribution stage.

The provision of accurate information to the public is highlighted as a key element of countermeasure implementation.

References

1. Konoplev, A.V., Borzilov, V.A., et al. (1992) Hydrological aspects of the radioactive contamination of water bodies following the Chernobyl accident, World Meteorological Organisation, Hydrological Report No 37, Hydrological aspects of Accidental Pollution of Water Bodies, Geneva, pp. 167–190.
2. Law of Ukraine (1992) On legal regime of the territory exposed to radioactive contamination as a result of Chernobyl accident.
3. The Water Code of Ukraine (1995).
4. Feasibility study for the renewal and modernisation of Kiev Vodokanal (2001) COWI, Final Programme Report, vols 1 and 2, January 2001.
5. Petimko, P.I., Tsarik, M.F., et al. (1997) Water supply in Kiev (in Ukrainian), Logos.
6. Law of Ukraine (1994) On sanitary and epidemiological protection of the population.
7. State sanitary norms and regulations on drinking water hygiene requirements of the quality of centralized water supply systems, No 383 of 23 December 1996 and No 136/1940 of 15 April 1997.
8. Novitsky, M.A. and Nikitin A.I. (1996) Prediction of radioactive contamination of water bodies after the Chernobyl accident, Radiation Protection Dosimetry 64 (1/2):29–32.
9. Zlobenko, B.P., Vovk, I.F., et al. (1989) Recommendation measures for prevention of wash-out radioactive contamination of surface ground out of the Pripyat water-meadow on the base of radiohydrogeological monitoring, Contribution No c-89 with Institute Ukrvodproekt of the State Committee of the Ukraine on Water Management (in Russian).
10. Zlobenko, B.P., Vovk, I.F., et al. (1991) Clay radiological barrier and countermeasure research related to the accident at the Chernobyl NPP, Proceeding of UIR Soviet Branch Seminar on the Radioecology and Countermeasures, Kiev, pp. 67–82.
11. Scherbak, A.V., Voitsekhovich, O.V. (1991) UHMI report Radiation monitoring of water bodies in 30-km zone of ChNPP and scientific substation of water protection measures on the base of investigation of migration processes in surface waters, Kiev (in Russian).
12. Voitsekhovitch, O.V., Nasvit O.I., et al. (1996) Present concept on water remedial activities on the areas contaminated by the Chernobyl accident, Health Physics 71 (1):19–28.
13. Voitsekhovitch, O.V. "Overview of Water Quality Management in the Areas Affected by the Chernobyl Radioactive Contamination", Radioecology and the Restoration of Radioactive-Contaminated Sites, Edited by F. Luykx and M.J. Frissel, NATO, Series 2. Environment Vol. 13, pp. 381–388. (1996).
14. Zlobenko, B.P., Vovk, I.F., et al. (1989) Study of sorbent materials for treatment of water, Contribution No 9–89 with Scientific and Industrial Association Pripyat of the Ministry of Nuclear Energy of the USSR (in Russian).
15. Fedorenko, Yu.G., Zlobenko, B.P., et al. (1992) Study dispersed and mineral structures of contaminated radionuclide soil, in: Radioisotopes in Ecological Researches, Kiev, Naukova Dumka, pp. 82–87 (in Russian).
16. Smith, J.T., Voitsekhovitch, O.V., et al. (2001) A critical review of measures to reduce radioactive doses drinking water and consumption of freshwater foodstuffs. Journal of Environmental Radioactivity 56 (1–2):11–32.

3.3 DRINKING WATER QUALITY MONITORING IN EMERGENCY SITUATIONS IN THE REPUBLIC OF ARMENIA

VARDAN TSERUNYAN
USAID's Sustainable Water Resources Management Project in Armenia, 48 Tigran Mets Ave., #6, Yerevan 0005, Republic of Armenia
Tel.: +374-10 55-06-53; e-mail: vtserunyan@yahoo.com
BARDUKH GABRIELYAN
Institute of Hydroecology and Ichthyology, Armenian National Academy of Sciences, Armenia

Abstract: The Republic of Armenia has developed a legal framework and regulative mechanisms to secure the quality of drinking water supply and population safety in normal and emergency situations. *Drinking Water Quality State Standards and Norms* (2002)[1] define the hygienic requirements, and regulate quality control of the drinking water produced and supplied to localities through centralized systems. With respect to monitoring of drinking water quality in the cities, both industrial and statutory control is implemented. Under *emergency conditions*, the Ministry of Health and Department for Emergency Situations define intensified regime for drinking water quality monitoring of the distribution networks and sources, according to the approved package plan. *Lake Sevan*, being the largest freshwater body in Armenia and in Southern Caucasus, is the unique strategic reserve of drinking water in the region. Overexploitation and pollution caused widespread changes throughout the entire lake basin. Uncontrolled abstraction of the lake's water brought about abrupt drop in water levels, resulting in eutrophication and water quality deterioration. Regular monitoring of the lake's water quality, aimed to control irreversible changes in Lake Sevan, is a *national priority* that will serve as a powerful tool for environmental impact assessment of any future hydrological engineering development in the country.

Keywords: drinking water quality control, emergency conditions, monitoring parameters, disaster zone, national priority, strategic reserve of drinking water

1. Quality Control of Drinking Water Supply

The Republic of Armenia has developed a legal framework and regulative mechanisms to safeguard the quality of drinking water supply and population safety in normal and emergency situations. The Republic of Armenia *Law on Securing Sanitary-Epidemiological Safety of Population* regulates the safety of drinking water supply. *Drinking Water Quality State Standards and Norms* (2002) define the hygienic requirements (health standards), and regulate quality control of the drinking water produced and supplied to localities through centralized systems. Its enforcement is authorized to the Ministry of Health State Hygiene and Anti-epidemic Inspectorate (Health Inspectorate), and its centers countrywide. Drinking water supplied to the cities should be safe in terms of epidemics and radiation, chemical composition, should have favorable organoleptic indicators (taste, odor, turbidity, etc.) and comply with the national health standards.

With respect to monitoring of drinking water quality in urban areas of the country, Industrial and statutory sanitary epidemiological control is implemented. In *emergency situations* or technical violations resulting (or which may result) in impairment of drinking water quality or water supply, the utility implementing control over the drinking water quality takes urgent measures to eliminate the problem. The utility alerts the Regional/District Centers of the Health

[1] Based on WHO Recommended Standards for Drinking Water (2000).

Y. Sharan et al. (eds.), Water Supply in Emergency Situations, 71–75.

Inspectorate on the noncompliance of drinking water quality with the health standards posing a risk to public health. Prohibition on drinking of tap water by population may follow, issued by the Chief Sanitary Doctor (MD) of the administrative area. If a decision is made to prohibit drinking of tap water, the utility (that runs the system) jointly with Regional/District Centers of the Health Inspectorate coordinate the actions to determine and eliminate the causes of impairment of drinking water quality. The population is notified by the local authorities and the Regional/District Centers of the Health Inspectorate on holding up the use of tap water for drinking needs.

In the city of Yerevan, with over one million population, drinking water quality is periodically monitored by the Chemical-Bacteriological Laboratory of the water supply and wastewater disposal utility (company). The laboratory functions according to the work plan approved by the Ministry of Health, and acts according to the regulations approved by the board of the company or utility. The laboratory serves the whole Yerevan city area and 22 rural (suburban) communities by conducting complete control or testing of chemical and bacteriological water quality parameters (quarterly) of the 15 springs or water mains that supply the city, whereas the 23 daily storage reservoirs – on a daily basis at the outlets (control of chlorination levels, i.e. residual chlorine, is checked every 2 h/day). In cases of noncompliance of chlorination levels to the standard (residual free chlorine [0.3–0.5 mg/l], residual combined chlorine [0.8–1.2 mg/l]), the laboratory notifies the administration of the reservoir. On a routine basis, the laboratory conducts complete analysis of water (organoleptic, ammonia or nitrogen, chlorine, and bacteriological) once a month.

Within the distribution networks, the industrial control over bacteriological and organoleptic indicators is conducted with a frequency defined in Table 1. Samples are taken from the water distribution conduits of streets that are located at the highest and dead sections, as well as residential service connection system (e.g. 62 characteristic sampling points in the city of Yerevan). At 36–37 sampling points of the city of Yerevan, organoleptic, ammonia or nitrogen, free chlorine, and bacteriological indicators are monitored daily.

TABLE 1. Requirements for sampling and analysis frequency of bacteriological and organoleptic indicators within the distribution networks

Number of serviced population (1,000 people)	Number of samples per month*
Up to 10	2
10–20	10
20–50	30
50–100	100
Over 100	100 +1 for each extra 5,000 people

**Not included* the number of samples taken for mandatory drinking water quality control after repairing or technical works on the network

For *sources of water uptake*, bacteriological/pathogenic, chemical and radiological parameters, and organoleptic properties are monitored on quarterly (groundwater source) or monthly basis (surface springs and reservoirs), at the points before entering the distribution networks. In emergency situations, the laboratory conducts "unscheduled" checks, and operatively reports on results to the utility and Health Inspectorate, to jointly undertake adequate measures (e.g. disinfection).

Under *emergency conditions*, the Ministry of Health and Department for Emergency Situations define an intensified regime for drinking water quality monitoring of the distribution networks and sources, according to the approved package plan. If a certain area, facing man-made or natural emergency situations, has been announced as a *disaster zone* by the Department for

Emergency Situations, then the latter has the authority to revise the water use permitting conditions and limits of maximum permissible discharges, provided they:

- Hold up elimination of the emergency situation.
- Impede provision of drinking water of adequate quantity and quality to population of the disaster zone.

1.1. LAKE SEVAN: STRATEGIC RESERVE OF DRINKING WATER

Lake Sevan, being the largest freshwater body in Armenia and in the whole South Caucasus region, is an enormous potential source of drinking water. Overexploitation and pollution caused widespread changes throughout the entire lake basin. Uncontrolled abstraction of the lake's water brought about abrupt dropdown in water levels (Fig. 1), resulting in eutrophication and water quality deterioration (Hovhannisyan, 1994).

In order to prevent the irreversible changes in the lake, the government has proclaimed an emergency situation for the Lake Sevan.

Lake Sevan is situated in the central part of the Republic of Armenia, about 1,900 m above sea level. The total area of the lake's basin is about 4,900 km^2, of which 3,650 km^2 constitutes the catchment area and approximately 1,250 km^2 constitutes the lake surface itself. The volume of the lake is 35.8 billion cubic meters. Twenty-nine rivers and streams are tributary to the lake, whereas the river Hrazdan is the only outflow from the lake. Lake Sevan is the largest freshwater lake in the region (Fig. 2), has a major contribution in formation of drinking water sources in the region (South Caucasus, Turkey, and Iran).

According to the state program, water quality is monitored routinely at six locations on Lake Sevan, at the depths of 5 and 20 m at each location, as well as from 24 locations on rivers flowing into lake (LeGore and Tserunyan, 2002). Most indicative water quality parameters and methods of determination are listed in Table 2.

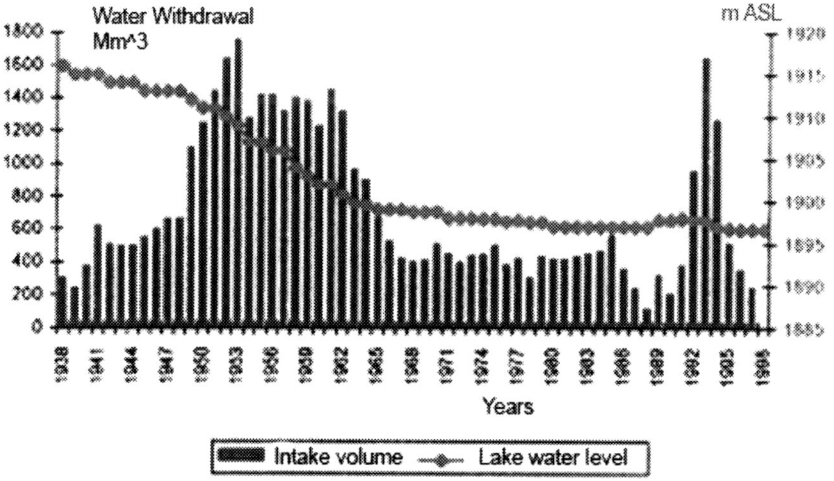

Figure 1. Water level variations of the Lake Sevan

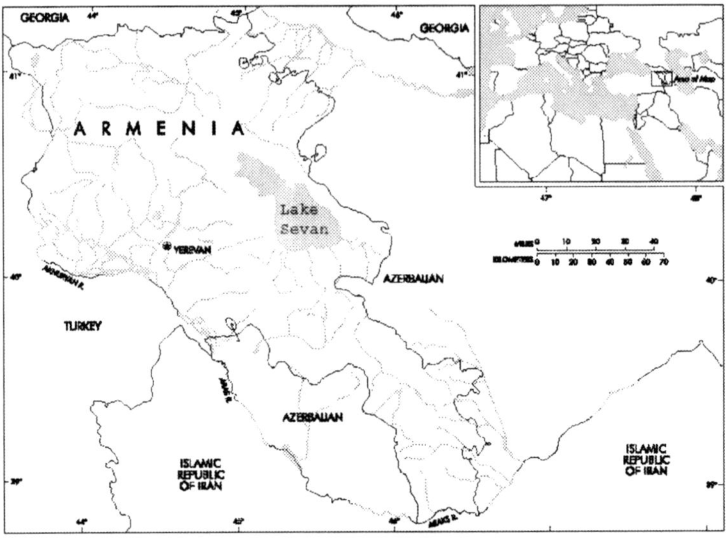

Figure 2. Map of the region

TABLE 2. Lake Sevan basin: routine water-quality monitoring parameters

Parameter	Analysis location		Analytical instrument or method
	Field	Lab	
Water temperature	X		TROLL 9000
pH	X		TROLL 9000
Dissolved oxygen	X		TROLL 9000
Total dissolved solids	X		TROLL 9000
Chloride – Cl^-	X		TROLL 9000
Ammonium – NH_4^+	X		TROLL 9000/Hach colorimeter
Nitrate – NO_3^-	X		TROLL 9000/Hach colorimeter
Nitrite – NO_2^-	X		Hach colorimeter
Sulfate – SO_4^{2-}	X		Hach colorimeter
BOD_5		X	Respirometric (HACH BODTrak)
COD		X	Spectrophotometric
Bicarbonate – HCO_3^-		X	Titration
Turbidity	X	X	TROLL 9000/turbidimeter
Oil and Grease		X	IR photometric
Phosphorus (as PO_4^{3-})		X	ICP-MS
Sodium – Na^+		X	ICP-MS
Potassium – K^+		X	ICP-MS
Total Iron – Fe		X	ICP-MS
Calcium – Ca^{2+}		X	ICP-MS
Magnesium – Mg^{2+}		X	ICP-MS
Mercury – Hg		X	ICP-MS
Lead – Pb		X	ICP-MS
Cadmium – Cd		X	ICP-MS
Chromium – Cr		X	ICP-MS
Nickel – Ni		X	ICP-MS
Copper – Cu		X	ICP-MS
Aluminum – Al		X	ICP-MS
Arsenic – As		X	ICP-MS
Boron – B		X	ICP-MS

Inductively coupled plasma mass spectrometer

Restoration of the lake's ecosystem and freshwater resources is a national priority for Armenia. Routine monitoring of the water quality, aimed to control irreversible changes in the Lake Sevan, has to serve as a powerful tool for environmental impact assessment of any future hydrological engineering developments and efficient management of resources in its basin (Hovhannisyan and Gabrielyan, 2000). These will bring about the improvement of the lake's water quality to meet the qualifying standards of strategic drinking water sources.

References

1. Hovhannisyan, R.H. (1994) The Sevan Lake: Yesterday, Today, Tomorrow. Armenian National Academy of Sciences, Yerevan.
2. LeGore, S.R. and Tserunyan, V.V. (2002) Lake Sevan – River Hrazdan Surface Water Resources Monitoring Strategy, ARD, Yerevan.
3. Hovhannisyan, R.H. and Gabrielian, B.K. (2000) ecological problems associated with biological resource use of Lake Sevan, Armenia. Ecological Engineering 16:175–180.

3.4 SECURE WATER SUPPLY AS A CENTRAL ELEMENT OF SUSTAINABLE WATER MANAGEMENT: SYSTEMS ANALYSIS APPROACH FOR THE GUNUNG SEWU REGION IN CENTRAL JAVA, INDONESIA

HELMUT LEHN*
Institute for Technology Assessment and Systems Analysis, Helmholtz Research Center Karlsruhe, Hermann-v-Helmholtz-Platz 1, D-76344 Eggenstein-Leopoldshafen, Germany, e-mail: lehn@itas.fzk.de

ANNE SCHILDT
Department of Urban and Environmental Sociology, Helmholtz-Center for Environmental Research UFZ Leipzig, Permoserstraße 15, D-04318 Leipzig, e-mail: anne.schildt@ufz.de

JÜRGEN KOPFMÜLLER
Institute for Technology Assessment and Systems Analysis, Helmholtz Research Center Karlsruhe, Hermann-v-Helmhotz-Platz 1, D-76344 Eggenstein-Leopoldshafen, Germany, e-mail: kopfmueller@itas.fzk.de

Abstract: Diversification and redundancy are important technical features to make infrastructure systems less vulnerable to natural and social (man-made) hazards. Fair access to basic natural and economic resources, public security and solidarity are important prerequisites for an adequate risk management in order to avoid catastrophes resulting from adverse events.Within an applied case study for the rural area of Gunung Sewu ("Land of the thousand hills") in Central Java, Indonesia, the possibilities of combining centralized and decentralized elements of a water supply and waste water treatment system will be assessed from the perspective of systems analysis.

Keywords: Indonesia, centralized water infrastructure, decentralized water infrastructure, natural hazards, resilience, reversibility, risk management, security, social hazards, sustainable development, vulnerability, water management

1. Introduction

Societal security and prosperity are highly dependent on the reliable functioning of the worlds' increasingly complex and interdependent infrastructures (energy systems, water supply and sewage treatment systems, information and communication systems, transportation, banking, etc.). Among these, freshwater supply is fundamental in almost all spheres of society and economy and necessarily has to be ensured in normal operation as well as in cases of emergency. The case study of Gunung Sewu in Central Java, Indonesia, demonstrates the consequences of a deficient or even missing water infrastructure for community life in remote rural regions under the adverse geological situation of karst and earthquake risk. Caught between already long-lasting problems with traditional water supply and new ones from modern pipe-bound systems, the designing of a solution feasible, stable, and culturally appropriate, is an enormous challenge. In this paper, we want to explain the special case of Gunung Sewu and discuss possible solutions under the aspect of a maximum effect in vulnerability limitation. *The main goal of that project will be to establish a sustainable water management which prevents – among others – the "secondary disaster" which means a lack of drinking water resulting from a primary disastrous event like an earthquake*. This article shall mainly focus on the systems analysis approach used. General considerations about technical standards or international rules concerning emergency water supply can be found, e.g. in Fader et al. (2002).

2. Vulnerability as a Function of Susceptibility, Resilience, and Resistance

The growing complexity and interdependency of infrastructure systems increase both their individual and common vulnerability. The resulting effects can have broad regional, national, and even global consequences. Vulnerability in this context is understood to be a function of both susceptibility to a certain event and the ability to react with resilience *or* to offer resistance

Y. Sharan et al. (eds.), Water Supply in Emergency Situations, 77–86.

(Klein, 2000). As the system's vulnerability is increased by growing susceptibility, it also decreeses according to the system's growing resilience or resistance.

Risk in this context is defined as the possibility or probability of an event's occurrence leading to considerable damage or a catastrophic impact. Accordingly, the term of "risk management" is defined as the sum of actions suitable for avoiding or minimizing both the event's probability and the resulting extent of damage.

For structuring risk management, it seems appropriate to distinguish between:

1. Natural hazards.
2. Natural hazards increased by human activities.
3. Man-made hazards.

In principle, there are two options for political and societal action:

(a) Avoiding the event's occurrence.
(b) Reducing or minimizing the event's impact as far as possible in order to prevent considerable damage or to avoid catastrophes.

Box 1. Hazard categories

Natural hazards such as earthquakes, volcanism, and tsunamis occur without human intervention. Nonetheless, human behavior can mitigate or adapt to these events and their effects through evasion or technical measures.
Natural hazards increased by human activities emerge from the interplay of natural systems and human activity. Examples are settlements in flood plains or water supply from karstic aquifers (see Chapter 4). Likewise, deforestation and incidental erosion and the consequences of the increasing number and intensity of natural events due to man-made climatic change fall within this category.
Man-made hazards include the threat to human life and health resulting from socioeconomic (poverty, hunger) and political events like war, political instability, and repression. Hazards from waste, sewage, emissions, or incidences on industrial sites, fires, and oil accidents are among this category, too.

In case of *natural hazards*, there are no options of preventing the event itself from occurring or of influencing its probability (e.g. earthquake). In such cases, measures of risk management are limited to the enhancement of resilience, i.e. the improvement of the society's ability to deal with the event successfully in terms of adaptation measures.

In case of *hazards increased by human activities* and especially *man-made (social) hazards*, options of influencing incidence rates as well as the extent of damage exist. The system's vulnerability can be reduced by a diminished susceptibility as well as by an increased resilience or resistance. Within these hazard categories, risk management does not only encompass strategies to avoid the occurrence of certain events which might lead to damages or catastrophes, but also strategies of adaptation to limit damages.

Concerning the predictability of a hazard and reaction time, in this paper a distinction is made between

- *Sudden events*, i.e. single events in the form of 'catastrophic disasters' (e.g. tornado).
- *Creeping threats* with long-lasting effects (e.g. increasing pollution of water – eventually leading to health risks).
- *Disaster cascades* triggered by an initial catastrophe (e.g. an earthquake which destroys the infrastructure providing drinking water, leading to an epidemic).

3. Secure Water Supply as an Element of Sustainable Water Management

Secure water supply and wastewater treatment include a permanent availability of certain water quantities of defined quality per capita within a defined area (catchment, supply area), as well as the collection and treatment of the resulting effluent. For this purpose, complex systems are needed that are usually composed of natural, economic, social, and technical subsystems. Their interplay is exemplified in Box 2.

Box 2. Interrelationship between natural, technical, economic, and social system components of water supply

Even under conditions of *normal operation*, a functional integration of different system components is a necessary precondition, e.g. for the supply of a residential quarter with potable water and safe sewage treatment. This does not only require sufficient natural resources and an operative technical infrastructure (water extraction and purification, distribution lines, sewage system, and treatment), but also an appropriate system for financing, construction, and maintenance, as well as the willingness of the user to accept, e.g. water meters and to pay an adequate price for the water used. A basic condition for the socioeconomic component is a system capable of gaining information on migration dynamics and, thus, the number of water users. As soon as one of these components does not work, the entire system cannot be operated satisfactory. This could lead to supply bottlenecks or hygienic risks for the affected population.
A *typical hazardous event*, for example, as a consequence of an earthquake, is the damage of the physical distribution infrastructure (water network, pipes, etc.). The loss of potable water supply typically leads to waterborne diseases, such as typhus and cholera. Such a "secondary disaster" requires increased efforts within the health care system (hospital capacities, drugs, personnel). Probably, the capacity of the health system has already been impaired as a direct consequence of the earthquake ("primary disaster"). In addition, there is the likeliness that the medical infrastructure itself may have suffered from damage. In this case, the breakdown of a central system component can lead to the failure of the entire system.

On the *supply side*, security may be threatened by natural hazards (climate changes, earthquakes, floods), man-made hazards (sabotage, terrorism), or by hazards amplified by human intervention (e.g. water supply from karstic aquifers – see Chapter 4).

On the *demand side*, vulnerability of supply and disposal systems is mainly affected by man-made (social) hazards like the following the number of people to be supplied is not adapted to the amount of water available; water demand fluctuates due to external conditions – resulting in large oscillations of the system's working load; consumers are not willing to pay for the use of water; parts of the technical infrastructure are damaged intentionally or even stolen; theft of water; no willingness to adjust the demand behavior in times of diminished water supply.

These examples show that discussions on security of water supply cannot be reduced only to infrastructural vulnerability, but also have to consider its operability.

The appropriateness of risk management tools depends on the assets available ("Access to Assets Model") (Hidajat and Voss, 2002). These assets include financial resources as well as available technologies and infrastructures, knowledge, and information, as well as social assets like institutions, informal structures, security, and solidarity. The power of resilience or resistance against risks correlates positively with disposable assets of the affected population. The success of adaptation and prevention measures is also highly dependent on people's willingness to handle risks. Access to these assets as far as possible will be important for the success of a risk management.

The multiplicity of affected subsystems, their complex interactions, and also the variety of possible tools for action require an interdisciplinary approach aiming at a comprehensive risk management and a transdisciplinary approach comprising scientific and nonscientific aspects, including cooperation with institutions, decision makers, and other stakeholders from the very beginning of the management process.

According to the considerations explained in Chapter 2, strategies of coping with emergencies are not only limited to adaptation to these, but also cover their prevention. As will be explained below, considerations of secure water supply in cases of emergency should be easy to integrate in strategies for sustainable freshwater management.

In accordance with the principles of the Brundtland Report (WCED, 1987), a sustainable use of water resources requires to meet the needs of the present generation without compromising the ability of future generations to meet their own needs. Concerning the topic of justice between different generations (intergenerational justice), the management of groundwater resources and the oceans is primarily important because of their long regeneration times. Concerning the justice within one generation (intragenerational justice), attention should be paid to the management of streams and rivers to meet the often diverging interests of upstream and downstream riparian owners (e.g. rivers Euphrates, Tigris, Jordan) (Lehn et al., 1996).

The postulate of justice in the field of water also implies that all people within a catchment area should have equal access to the same amount of affordable water of good quality; no one should be excluded from water use, not the least in order to diminish the probability of sabotage or terror as a result of unfair treatment and, thus, man-made (social) hazards.

As far as technical options are concerned, strategies of reversibility and the ability of adaptation to changing general conditions are highly relevant in the field of sustainable resource management. They can also be useful in cases of emergencies. The water infrastructure developed in Europe during the 19th century and still applied, is almost completely based on options of centralized systems: huge supply and disposal networks with few, but large waterworks and sewage treatment plants.

The concept of widely ramified networks combined with the physical water pipe and sewer system itself in practice proves to be almost irreversible. On the one hand, this is a fact from the economic point of view. Reconstruction measures are steadily being conducted on different system components. Consequently, the system can never be classified as entirely depreciated and worth renewing at once. On the other hand, people seem to consider the transport of human excreta with water as "law of nature," consequently people often do not accept dry sanitation systems. During the last two decades, more decentralized systems for water supply and wastewater treatment have been developed in Central Europe, which are often based on the principle of partial flow treatment and frequently use membrane technology (Lehn, 2004). Transformation of a rather centralized system into a more decentralized system requires the parallel operation of two systems for a transition period, which would be rather expensive. The effect is a "lock in" situation in Central Europe concerning the possibilities of system alteration. If new decentralized or semidecentralized system options are implemented at all, this is the case in newly built quarters or remote regions without any connection to the existing central supply and return network.

Thus, in developing and newly industrialized countries without any water and sewage infrastructure, the preconditions for the establishment of such new system options are much better, if appropriate technologies are applied.

4. The Case of Gunung Sewu in Central Java (Indonesia)

The region of *Gunung Sewu* is situated in a tropical karstic landscape in Central Java, which is affected by a monsoon climate. The term *karst* describes a special type of rock weathering in limestone and gypsum. Typical features of this phenomenon are caves as well as clefts and

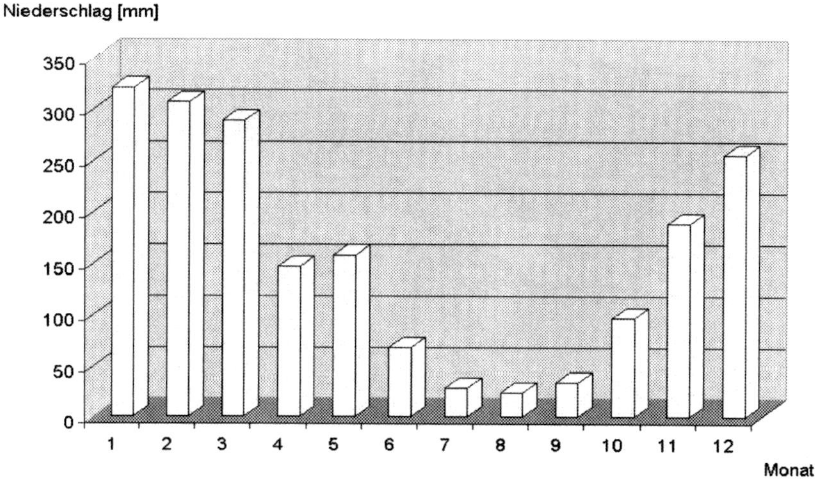

Figure 1. Precipitation (mm) near the town of Wonosari, Gunung Kidul, Central Java (time series 1952–1955, 1968–1981). (From University of Karlsruhe, Institute of Aquatic Environmental Engineering, http://www.hoehlenbewirtschaftung.de/English/English.html.)

cracks formed by permanent water-driven rock dissolution. This results in missing surface waters due to strong and rapid seepage and the accruement of subsurface water flows (Scholz et al., 2004).

Precipitation amounts up to 1,800–2,000 mm per annum, the major part of it falls in the rainy season. Even though there is a relatively high precipitation, seasonal disparities are typical (see Fig. 1).

The bulk of rainwater immediately vanishes through clefts and is currently not detained on the surface. Therefore, the availability of water for use is very limited at present. Particularly with regard to its lack of water and the limited tillage options, the region of *Gunung Sewu* is being called the "Poorhouse of Java" (Figs. 2 and 3).

Figure 2. The tower karst region of Gunung Sewu in the rainy season. (From Scholz et al., 2004.)

Figure 3. The tower karst region of Gunung Sewu in the dry season. (From Scholz et al., 2004.)

Traditionally, for water supply the population had predominantly relied on small shallow ponds – so called telagas. For most of the telagas, had dried out by the middle of the dry season, people partially had to cover enormous distances to the next telagas that was still filled.

At the beginning of the 1970s, governmental support programs initiated the building of rainwater cisterns. Population growth led to an increasing water demand. In 1981, the *first Bribin water scheme*, a water pipe network fed with water from the underground, *Bribin* – karst aquifer (Fig. 4), was introduced, financed by UNICEF and the Indonesian Ministry of

Figure 4. Underground dam of the Bribin scheme built in 1981. (From University of Karlsruhe, University of Karlsruhe, Institute of Aquatic Environmental Engineering, http://www.hoehlenbewirtschaftung.de/English/English.html.)

Health. Technical problems like broken pumps and pipes, as well as a lack of pressure to run the system correctly and completely led to enduring problems in water supply. Villages situated far from the Bribin aquifer hardly got any water from the water system, since the pressure in the pipes was very low.

At the beginning of the 1990s, Indonesian president Habibie ordered to construct additional water pipes in the region of Gunung Sewu. Nonetheless, the existing technical problems remained unsolved. Therefore, many people still had to use other water sources in addition to the *Bribin* network. The water supply system was not reliable, neither in terms of the availability, nor in terms of quality and general dependability.

Since the year 2000, the German–Indonesian project "Water Resources Management of an Underground River in a Karst Area" has been devoted to analyses of management options for making available a greater amount of water from the Bribin aquifer. Currently, a new subsurface dam is being built to seal off the Bribin cave and dam up the water that otherwise would flow to the sea unused. The designed delivery volume of 400,000 m^3 would allow a sufficient supply (50 l/day person) for some 22,000 people. If all 75,000 people in Gunung Sewu connected to the Bribin water pipe would use the water, an amount of about 15 l/day could be provided by the pipe network. The use of pumps as turbines (PAT) has proved to be the most appropriate technology in terms of robustness, cost minimization, availability, and maintenance (see Fig. 5). In the next project step, the existing pipe network will have to be renovated or rebuilt.

A sufficient pipe-based water supply for a great part of the population is intended to improve the living conditions and especially the development of regional economy, e.g. by increased agricultural irrigation or home processing of agricultural products for trade on regional markets. Taking into account different kinds of possible hazards (see Box 1) the chosen centralized system is or might be exposed to, the following question arises from the point of view of technology assessment and systems analysis:

Figure 5. Model of the new dam with PAT technology. (From University of Karlsruhe, University of Karlsruhe, Institute of Aquatic Environmental Engineering, http://www.hoehlenbewirtschaftung.de/English/English.html.)

How could the water supply system in this region be designed to minimise the vulnerability regarding the three major types of hazards?

Finding the answer to this question shall be one important topic within a research project that has been applied for a 5-year term from 2007 to 2011, and which will be financed by the German Federal Ministry of Education and Research. This project will focus on looking for strategies to provide secure water supply in the dry season as well as to prevent so-called secondary disasters (e.g. epidemics) resulting from e.g. an earthquake, by interrupting the related disaster cascade (see Chapter 2).

As far as *natural hazards* are concerned, the region is situated quite close to one of the world's most active volcanoes, the Merapi (Hidajat and Voss, 2002). Moreover, the probability of earthquakes has to be rated as high, for Java is part of the circumpacific "Ring of Fire" (Fig. 6). As the outbreak of the Merapi and, in particular, the earthquake in 2006 have shown, there is a high risk of damages to the water infrastructure. Thus, broken pumps and pipes might paralyse the regions' water supply. Additionally, streets might become impassable with the consequence that external water supply by tank trucks as well as the repair of water infrastructure could be impossible for days or weeks.

Within the centralized water supply system of the new Bribin network, redundancy can be implemented by the parallel running of four pumps and by linking the pipe network to other networks outside the Bribin scheme. Pipes above ground are easier to maintain and to repair compared to underground pipes, but they are more susceptible in the case of an earthquake. Restoration and extension of existing cisterns and construction of new ones could strengthen the role of decentralized infrastructure and make the entire system more resilient by increased diversity.

As the pipes pass above ground and air temperature is high, microorganisms are growing in the pipes on the way from the water source to the tap. Therefore, it is considered to operate a freshwater treatment near the tap and not near the source. At present, people are used to boil water before drinking. In the future, the use of membranes or ultraviolet (UV) light might be an option for disinfection (Martini, 2005). In the case of freshwater treatment near the tap, it

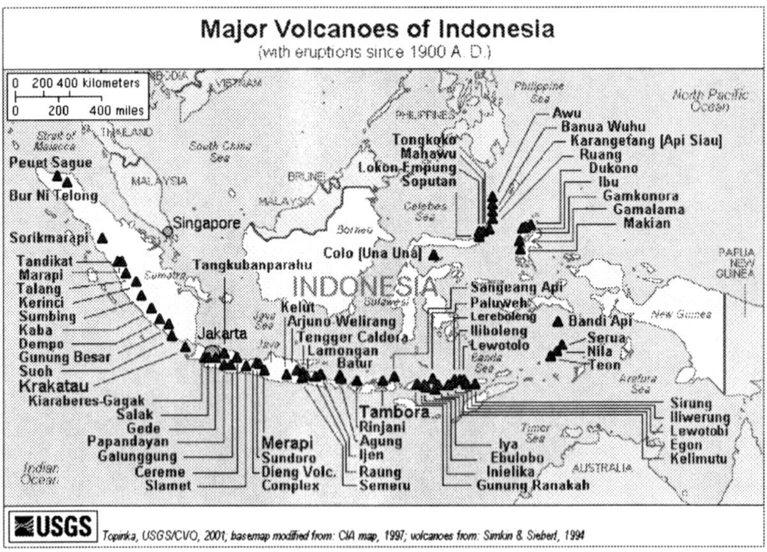

Figure 6. Java – a part of the circumpacific "Ring of Fire" (USGS) (http://vulcan.wr.usgs.gov/Volcanoes/Indonesia/Maps/map_indonesia_volcanoes.html)

would be possible to use the water from cisterns for drinking purposes as well. Thus, the cisterns could be filled both with Bribin water and rainwater. In the rainy season, people are expected to prefer rainwater from the cisterns because it is almost free of charge and because of the lower water hardness compared to calcareous Bribin water. This means a certain risk concerning the continuous operation of the water infrastructure. Appropriate regulations for the supply of water (flat rates or price for the supplied volume of water) are considered to be a prerequisite for the management of a centralized pipe network under seasonally inconstant working conditions.

Among the natural hazards, also El Nino plays an important role for Java. Its occurrence intensifies the problem of water shortages, as it significantly extends the duration of the dry season. A well-working pipe network should be able to satisfy the basic demand of people in Gunung Sewu in case that cisterns and telagas will dry out.

With respect to the category of *natural hazards increased by human activities*, the special problem of pollution of karstic aquifers seems to be the most important risk. This applies to the probability of its occurrence and the extent of the possible damage. As the precipitation water infiltrates very quickly along clefts and cracks, there is hardly any natural filtering effect. Consequently, the karstic Bribin aquifer is highly vulnerable with respect to water quality, especially due to feces, effluents, oil, and fertilizers. Political measures according to the precautionary principle should aim at installing appropriate land use practices, providing modern facilities for maintenance and repair of the quickly increasing number of motorbikes, trucks, and buses, and last, but not least, designing and installing secure systems for sewage storage and treatment.

Because of the rural character of Gunung Sewu, decentralized (septic tanks) or semicentralized options (e.g. anaerobic digesters with biogas production) could play a major role for sustainable sewage treatment. Semicentralized solutions should be designed in a way that allows for later treatment of additional sewage from scattered sewage collection units nearby.

Concerning *man-made hazards*, attention should be paid to a fair contract between water suppliers and water users. This includes the price, quality, and steadiness of water services.

Figure 7. Demounted water pipe at the foot of a cistern in Gunung Sewu. (Courtesy of H. Lehn.)

For example, prevalently waterless pipes make people use these pipes for other purposes (e.g. protection of power supply lines) and parts of them are stolen (Fig. 7).

For Gunung Sewu, it is reported that it takes the water authorities a long time to repair broken pipes or water pumps. For the people affected, it might be a considerable risk to rely on one centralized water supply system only. By the combination of centralized and decentralized water infrastructure, susceptibility of people to potential improper business practice (e.g. monopolized structures, corruption) would be clearly reduced.

Technologies for decentralized or semicentralized water supply and wastewater treatment are available in the meantime. Their prudential combination and linkage to centralized systems now has to be analysed and adapted to the regional socioeconomic conditions in order to achieve a secure and sustainable water management. Implementing these ideas in practice might open up a real chance for Gunung Sewu to overcome its seasonal *and* structural water shortages and to make it less vulnerable to most kinds of hazards.

References

1. Fader, H.J., Maier, M., Gruber M., Parandian, A., and Maier, D. (2002) Internationale Anforderungen an die stationäre und mobile Notwasserversorgung, in: AWBR (ed.) Arbeitsgemeinschaft Wasserwerke Bodensee-Rhein, Jahresbericht, Stuttgart, pp. 139–201.
2. Hidajat, R. and Voss, H. (2002) Ein Konzept zur Vulnerabilität verdeutlicht am Hochrisikovulkan Merapi auf Java, in: Tetzlaff, G., Trautmann, T.U., Radtke, S. (eds) Zweites Forum Katastrophenvorsorge Extreme Naturereignisse – Folgen, Vorsorge, Werkzeuge. Bonn, Leipzig.
3. Klein, R.J.T. (2002) Costal Vulnerability, Resilience and Adaptation to Climate Change. An Interdisciplinary Approach, Kumulative dissertation, Universität Kiel.
4. Lehn, H. (2004) Mit Vielfalt zum Erfolg – Globale Veränderungen erfordern angepasste Technologien, ZfK – Zeitung für kommunale Wirtschaft 50 (10):18.
5. Lehn, H., Steiner, M., and Mohr, H. (1996) Wasser, die elementare Ressource. Leitlinien einer nachhaltigen Nutzung, Springer, Berlin/Heidelberg/New York.
6. Martini, F. (2005) Vom Bakteriencocktail zum Lebensspender, in: Siemens, A.G. (ed.) Pictures of the Future, Siemens, Munich, pp. 69–72.
7. Scholz, U., Unger B., and Lux, T. (2004) Erschließung und Bewirtschaftung unterirdischer Karstfließgewässer in Mitteljava, Indonesien, Sozioökonomische Analyse der potentiellen Wassernutzer, Abschlussbericht, Karlsruhe/Gießen.
8. WCED (World Commission on Environment and Development) (1987) Our Common Future. The Brundtland Report, Oxford University Press, Oxford.

3.5 THE ANTHROPOGENIC POLLUTION OF WATER BASINS: THE ISSUE OF WATER SUPPLY SYSTEM SAFETY

[1]CHRISTINA HAKOPIAN AND DIANA HAKOBYAN
Yerevan State University, Yerevan, Armenia
Yerevan State Linguistic University, Yerevan, Armenia

Abstract: Man has used natural resources and influenced nature from ancient times. Since then the ecological balance has been destroyed to the extent that it is necessary to reconsider the "man–nature" interaction. In the "man–nature" system the decisive role is given to the relations "man–water resources."

Before the 20th century, the process of maintaining the balance in water basins was not violated, and water basins "managed" to self-purify. At present, they cannot accommodate the anthropogenic load, and the issue of use and conservation of water resources and water ecosystems has become very complicated and urgent.

The anthropogenic impact causes harm to both quantity and quality of water resources. However, water pollution harms only the environment. It has a boomerang effect and serves against man as well. The effects of water pollution on man can be regarded as direct and indirect. To conserve water resources from pollution and exhaustion under conditions of intensification of water use, a series of activities, which will ensure the appropriate condition of water bodies in accordance with the present legislation on water, are to be applied.

Keywords: anthropogenic pollution, self-purification, anthropogenic catastrophes, water quality monitoring, eutrophication

1. Introduction

Man has used natural resources and influenced nature from ancient times, when his activities were collection and hunting. During centuries the human impact on nature has changed, becoming more intense. Yet, before the 20th century, this impact was relatively insignificant, it did not affect the natural balance, and small changes like pollution could be "recovered" by the law of self-rehabilitation and self-purification.

At the beginning of the 20th century, academician Vernadskij compared the human activities to geological forces. However, if natural disasters, such as volcanoes, earthquakes, tsunami, could be "overcome" by nature through laws of self-regulation and self-rehabilitation, nature cannot cope with "anthropogenic catastrophes" (Hakopian Ch., 2002).

At present, the ecological balance has been destroyed to the extent that it is necessary to reconsider the "man–nature" interaction. Numerous examples show that humanity has reached the level when it needs to thoroughly analyze the fruits of its activity of the past, and act very carefully in future. The way of production, as well as use of natural resources must be revised; to do this, the regulations of the "man–nature" interaction must be established.

[1] To whom correspondence should be addressed. Christina Hakopian, Department of Physical Geography, Yerevan State University, 1, Alek Manoukian Street, Yerevan, 0025, Armenia; e-mail: chakopian@ysu.am

Y. Sharan et al. (eds.), Water Supply in Emergency Situations, 87–90.

2. Water Resources Self-Purification

In the "man–nature" the decisive role is given to the relations "man–water resources." Man has used water resources for ages and allowed the flow of "used" (sewage) waters into the water basin. Before the 20th century, the process of balance in water basins was not violated, and water basins "managed" to self-purify. At present, they cannot accommodate the anthropogenic load, and the issue of use and conservation of water resources and water ecosystems has become very complicated and urgent. Today, the opinion of inexhaustibility and revival of water resources on the planet is under question.

The anthropogenic impact causes harm to both quantity and quality of water resources. Water quality depends on natural and anthropogenic factors. As a result of intense use of water resources, changes have occurred not only in the quantity of water, fit for various branches of economy, but also in the water balance, the hydrological, hydrochemical, and hydrobiological regimes of the water body, as well as in the water quality. While being sources of water supply, rivers and lakes at the same time receive household, industrial, and agricultural wastewater flows. Today, it is hard to find a water basin with a natural hydrological regime and chemical composition, a basin which has not been utilized by man. Examining the degree of pollution of any basin in question, we often encounter the problem that not a single, but several factors of economic activity, such as water use for household and industrial needs, sewage water, urbanization, reservoir construction, irrigation and reclamation of arid lands, drainage, agro-melioration works, and so on, have been the causes of serious quantitative and qualitative changes of water resources (Vladimirov et al., 1991).

According to data from the Armenian Sanitation-Epidemiological Service, one fourth of utility pipelines and one third of nondepartmental supply of water are without sufficient purification, and the running water does not comply with hygiene requirements as shown by chemical or bacteriological indicators, bringing about high levels of infectious and noninfectious diseases (Avagian and Shirokov, 1994), especially in developing countries.

The natural water basins undergo pollution, processes of eutrophication, and thermophication, and water composition changes, which bring about the deterioration of water quality for water users and inhabitants of the basins. Anthropogenic pollution, eutrophication, and thermophication of water basins bring about quantitative and qualitative changes of physical–chemical quality of water.

The anthropogenically caused eutrophication, i.e. the increase of trophia level in water basins, forms as a result of excessive input of biogenes (nitrogen, phosphorus) under certain conditions.

Anthropogenic thermophication of water basins, i.e. change of temperature regime, is caused by pouring heated water from industries and, mainly, thermal power stations into water basins violates its thermal, hydrochemical, and hydrobiological regimes.

3. The Boomerang Effect of Water Pollution

Water pollution, however, harms not only the environment. It has a boomerang effect and serves against man as well. The effects of the water pollution on man can be regarded as directly through drinking water supply and indirectly through irrigation (Fig. 1). Both of them, certainly, cause larger damage in developing countries where due to financial and technical problems it is difficult to undertake necessary preventive or water treatment measures and policies. Actions such as renovation of water supply systems, change of pipelines, and installation of water filters are too costly for many of the developing countries and they actually rely on appealing for caution by the population or by merely cutting off water supply for the period needed to repair the existing problem. These means are not effective and bear consequences of disease and human casualties.

The indirect channel of harm by water pollution is related to irrigation. Polluted water damages the soil and, hence, the crops, which, containing the pollutants may affect human health to different degrees.

Such cases may take place in developed countries as well, which means that the industrial and economic developments are not on the equal scale with water supply system safety means. Thus, man-made chemicals and other pollutants related to human activity can and should be best controlled before the initial stage of pollution, i.e. though its prevention.

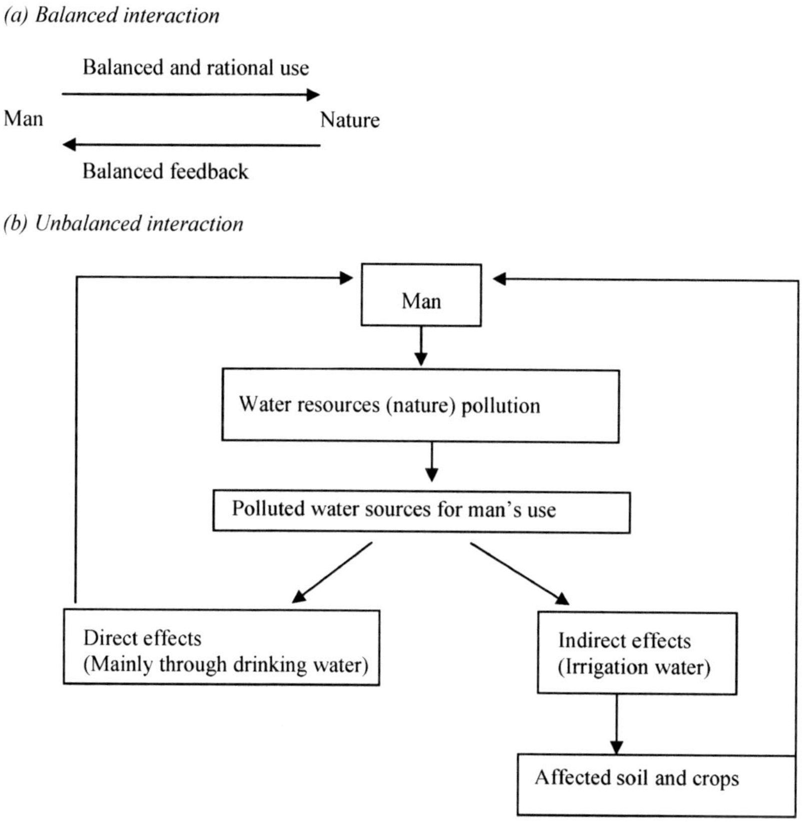

Figure 1. "Man–nature" interaction

4. Conclusions

Summarizing the analysis, we came to the following conclusions. To conserve water resources from pollution and exhaustion under conditions of intensification of water use, a series of activities, which will ensure the appropriate condition of water bodies in accordance with the present legislation on water, are to be applied. To implement these activities, a number of scientific, technical, and financial issues must be settled. At first, norm setting of water quality, i.e. the exact criteria for water use for any given water user must be defined. The quantity of sewage water, released into water basins, must be reduced by means of technological processes, i.e. modernization, which require considerable investments. Furthermore, the processes of self-purification in every single water basin must be studied, and strict control over the process of purification of sewage waters flowing into water basins is to be established. Besides these activities, one must consider the impact of every single factor and of all the factors combined on water quality for water economy planning and regulation of water quality.

Only a complex, systemic approach to the solution of this issue will allow us to settle the issue of rational use and conservation of water resources, to a certain extent.

To solve this issue we suggest to:

1. Establish and use a system of monitoring water quality.
2. Examine the possibility of changing the system of water supply and water pipes.
3. Improve the technology of water supply with consideration of respective state norms.

However, it must be stressed that realization of these suggestions needs significant investments, which is a serious issue in developing countries.

References

1. Hakopian, Ch. (2002) The consequences of the global anthropogenic changes in dry mountainous regions. Abstracts of the 8th International Conference on Understanding Future Dryland Environmental Changes from Past Dynamic, organized by Yard University with cooperation of UNESCO and the University of Sheffield, Yazd, I.R. of Iran, p. 20.
2. Avagian, A.B. and Shirokov, V.M. (1994) The rational use and conservation of water resources, Ekaterinburg, 320 (in Russian).
3. Vladimirov, A.M., Ljakhin, Yu.I., Matveev, L.T., and Orlov, V.G. (1991) Conservation of the environment, Leningrad, 424 (in Russian).

CHAPTER IV

DETECTION, MONITORING, AND WARNING

4.1 WATER SECURITY: OVERVIEW ON TECHNOLOGIES AND METHODS FOR RAPID DETECTION OF TERROR EVENTS

JEAN NEGREANU
The Interdisciplinary Center for Technology Analysis and Forecasting. Tel Aviv University, Tel Aviv, Israel

Abstract: After September 11, and the anthrax letters attack on government and press institutions in the USA in 2001 (Jernigan et al., 2002), the improvement of the water supplying systems security against terrorist attacks became a primary goal of water suppliers in many countries.

Due to the high complexity of the systems, and the innumerable possible points of contaminant insertion, complete prevention of all possible terror attacks (chemical, biological, or radiological) on modern drinking water supplying systems, practically, seems to be an impossible goal. For example, in the USA there are about 170,000 water systems, with about 8,100 very large systems that serve 90% of the population who get water from a community water system (Interim Voluntary Guidelines for Designing an Online Contaminant Monitoring System, 2004).

The prevailing approach to the problem of drinking water contamination is based on the implementation of surveillance measures and technologies for "risk reduction" such as improvement of physical security measures of critical assets (high-potential vulnerability to attacks), installation of online contaminant monitoring systems (OCMS) with capabilities to detect and warn in real time on relevant contaminants, as part of standard operating procedures for quality control (QC) and supervisory control and data acquisition (SCADA) systems.

Ideal detectors (OCMS) should be able to detect and warn automatically in cases that chemical contaminants are detected, at concentrations greater than the maximum allowable concentration in water (EPA list of contaminants whose maximum allowable concentrations in water are regulated, CDC list of biological contaminants, or radiological contaminants).

The nature of the threat posed by terror contamination of water delivery systems, and the short timetable (minutes to 1 h) in which effective steps can be taken to minimize the numbers of affected people, and the damage to the infrastructure is such, that even for risk reduction solutions, great numbers of online OCMSs have to be operated continuously.

Building and operating continuously such systems of hundreds (and possibly even thousands) of OCMSs, is problematic and very expensive, so that practically only few from the potential technologies of detection which are commonly used in laboratories (chemical and biological/clinical) are appropriate for monitoring contaminants in water supplying systems.

The main practical approach to the issue of online monitoring systems of terror contaminations of water is based today on development and improvement of known technologies for QC monitoring, using measurement of "surrogate parameters," for real time detection of an unusual/suspected terror event, and as a trigger for immediate actions. The final identification and validation of the contamination event usually can be done later, using normal laboratory techniques.

Y. Sharan et al. (eds.), Water Supply in Emergency Situations, 93–100.

Keywords: water security, rapid detection, risk reduction, OCMS, contamination scenarios, surrogates parameters

1. Terror Attacks on Water Supply Systems: Contamination Scenarios

Terror attacks on drinking water supply systems can be performed in many different ways. For example, chemical, radiological, or biological contaminants can be dumped into finished water storage tanks, injected into pipes under pressure through bleeder valves. or can be introduced under pressure into fire hydrants or taps (Drinking Supply: Terrorists had eyes on water, 2004). The result of a contaminant insertion is a contaminant "pulse" whose duration and concentration profile vary with the attack scenario and the properties of the water infrastructure.

Insertion of contaminants to water storage would result in long-lasting contamination, usually lasting several days, and reaching peek concentrations somewhere in the middle of the period. The effect of the contamination on the population can range from immediate illness and death in some cases, to usually less drastic results, depending on the nature of the contaminant and its concentration (potential contaminants discussed in the following paragraph). Following such an attack, utilities will have to deal with decontamination of water and the polluted infrastructures (pipes, valves, pumps, etc.), and with emergency water supply to the population in the affected area – until normal supply is restored.

2. Potential Contaminants

There is an enormous variety of potential water contaminants which can harm the consumers: chemicals, biological pathogens and toxins, and radioactive materials. Many of them are possible candidates for terrorists' use.

In such cases, the most probable scenarios of terrorist attacks include not only use of some of the more toxic materials known such as, biological and chemical warfare agents, but also agents from a much larger list of materials, based on their availability on the open market. For example, many of the water miscible pesticides in current use by farmers are high-risk threat agents in scenarios of terror attacks on potable water systems.

For a detailed discussion on the subject see the article *The Threat to Water Supply Systems* by Sharan, and references such as Burrows and Renner (1999); EPA list of contaminants whose maximum allowable concentration in water are regulated, and ATSDR database on toxic chemicals.

3. Detection Technologies

The basic goal of detection systems and technologies is to discover contamination events in (near) real time, and to give information about the location of the pollutants within the water supplying system. The main factors that are taken in consideration when comparing different detection technologies for their practical suitability to be employed in the field are:

- Reaction time (time between exposure to signal).
- Sensitivity.
- Reliability of detection (number of false alarms).
- Price of the system (acquisition, operation, maintenance, and life cycle).
- Suitability with existing technologies and operational procedures.

Ideal detector technologies of water contaminants should detect in real time and warn automatically when chemical contaminants are detected at concentrations greater than the maximum allowable concentration in water (http://www.epa.gov/safewater/mcl.html#mcls), or when biological or radiological contaminants (http://www.bt.cdc.gov/agent/agentlist.asp; http://www.cfsan.fda.gov/~mow/intro.html) are found in the water.

For the time being, the only viable approach to the issue of (near) real-time detection of water system terror attacks may be through development and implementation of online contaminant monitoring systems (OCMS) and technologies capable to detect and warn in real time, as part of standard operating procedures for QC.

Because today there are no available, online instruments able to detect most of the possible contaminants species, at needed sensitivities and timetables, the leading approach is to measure surrogate parameters as a means for indirect detection of the contaminants. Precise identification of most of the relevant species of contaminants is possible in expert analytical laboratories.

Table 1 (from Interim Voluntary Guidelines for Designing an Online Contaminant Monitoring System, 2004) shows some potential surrogates that might indicate the presence and concentration of relevant contaminants.

TABLE 1. Potential surrogates to monitor in detecting contaminants

Chemical surrogates	Microbiological surrogates	Toxin surrogates	Radiological surrogates
pH	Toxicity indicators	Total organic carbon	Alpha
Turbidity	Turbidity	Biomonitors	Beta
Total organic carbon	Phosphate	Toxicity indicators	Gamma
Chlorine residual	TOC		Toxicity indicators
Conductivity	Nitrate, nitrite		
Dissolved oxygen	Chlorine residual		
Nitrate, nitrite	Multiangle light scattering		
Phosphate	Fluorometry		
Oxidation reduction potential	Biomonitors		
$UV_{254/280}$	Biological oxygen demand		

[47] USGS National Water Quality Lab Newsletter, July 1999 for a discussion of the utility of UV absorption measurements at these two wavelengths. See also Eaton, A.D., Clesceri, L.S., and Greenberg, A.E., eds., 1995, Standard methods for the examination of water and wastewater. Washington, D.C., American Public Health Association. American Water Works Association, and Water Environment Federation, 19th edition.

One practical approach for implementation of surrogate measurements as an instrument for online (real-time) detection of pollutants in the water, may be through development of empirical sets of fingerprints, based on responses of sets of detectors to contaminants in water samples, taken from the specific water supplying systems, in parallel with measurement of good baselines of the these parameters over long periods. In such way, significant changes in the water quality, such as expected in many terror attacks scenarios should be detected.

4. Online Contamination Monitoring Systems of Surrogates Parameters: Instruments and Technologies

From the many OCMS technologies and instruments available today each water supplier will have to choose the most appropriate ones, based on the local needs and characteristics of the water systems, while taking in account practical issues as price for life cycle, suitability to existing QC systems and operational procedures, reliability, etc. Two updated sources of information on OCMS instruments and technologies are EPA publications (www.epa.gov/etv/verifications-index.html; www.epa.gov/safewater/watersecurity/guide;).

Examples of some leading candidate of surrogate parameters and instruments for indirect detection of water contaminants are elaborated below:

4.1. RESIDUAL CHLORINE

Residual chlorine is one of the most sensitive indicator parameters in water distribution system monitoring (Fig. 1). All water distribution systems use residual chlorine monitoring as part of their Safe Drinking Water Act (SDWA) procedures. Many chemical and biological contaminants combine with chlorine in water, lowering residual chlorine values significantly. Using empirical correlations between changes observed in residual chloride values, as part of a suite of surrogate parameters, it is one of the practical approaches to indirect (online) detection of natural contaminants and terror events in water (more information see at http: www.epa.gov/safewater/water).

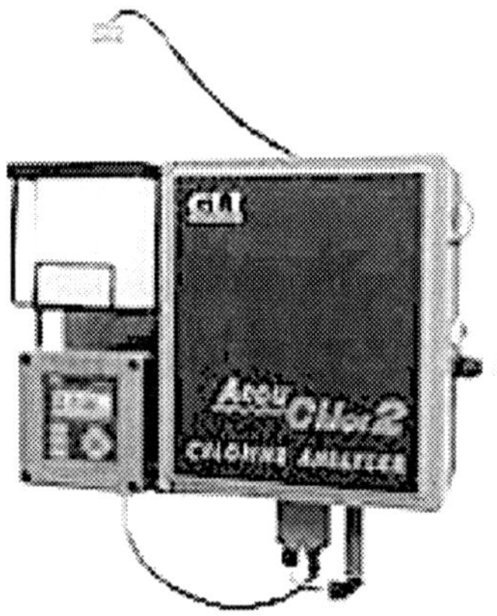

Turbidity meters

Figure 1. Residual chlorine measurement system, model AccuChlor 2/AC2000P1A1N, Hach. (From www.hach.com.)

Particulate contaminants as microorganisms or particulate chemicals may be detected by measurement of turbidity, and other optical properties of water (Fig. 2). Turbidity is defined as a parameter of the contaminants that causes light to be scattered and absorbed. Multiangle laser scattering (MALS) and motion analysis to determine the nature and amount of bacteria in water samples is one of the promising technologies for online detection of microbiological contaminations of water (more information see at www.epa.gov/etv/pdfs/vrvs/01_vr_abb.pdf and www.shu.ac.uk/scis/artificial_intelligence/biospeckle.html).

Figure 2. Turbidity system, model 4670 series, ABB instrumentation. (From www.epa.gov/etv/pdfs/vrvs/01_vr_abb.pdf.)

4.2. TOTAL ORGANIC CARBON

Total organic carbon (TOC) measurements are another commonly used method for measurement of carbon content of dissolved and particulate organic matter in water. Many water utilities

pH meters and conductivity meters

Figure 3. Continuous low-temperature online TOC/TC analyzer, model EZTOC, ISCO Inc. (From http://www.isco.com.)

use TOC measurements for raw water quality and efficiency of water treatment processes. Online TOC analyzers (Fig. 3) are designed to operate continuously and thus are adequate for continuous surveillance of water quality and as one of the possible surrogate online indicators of water pollution (terror) events (more information see at http://www.epa.gov/safewater/watersecurity/guide/chemicalsensortotalorganiccarbonanalyzer.html).

The pH and conductivity are some of the most useful parameters indicating changes in the ionic constitution of water. Combined with other surrogate parameters, they can be useful indicators of contamination events (Fig. 4).

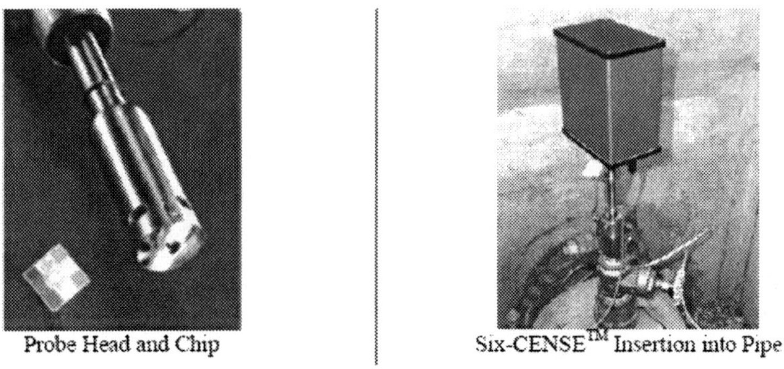

Figure 4. Six-CENSE multiparameter in-line sensor, Dascore. Measures: chlorine, monochloreamine or dissolved oxygen, pH, conductivity, oxidation–reduction potential, and temperature. (From http://www.dascore.com/sixcense.htm.)

4.3. PHOSPHATE, NITRITE, AND NITRATE MONITORING

Applying phosphate, nitrite, and nitrate continuous monitoring to drinking water is another option to indirectly determine microbiological contamination in the distribution system (Fig. 5). Nitrogen is a component of prokaryotic cells (Rittman and McCarty, 2001). Phosphorus is a required element for bacterial growth and it is found in nucleic acids, proteins, and many other intracellular materials. When cells are lysed, phosphate is released. In order to apply this technology for detection of biological and of some chemical contaminants in water, the water have to be sonicated (homogenized), before passing through the monitors.

Figure 5. Series 4 multiparameter water quality monitoring Sondes, Hydrolab. Measures: ammonium, chloride, conductivity, dissolved oxygen, nitrate, pH/reference, pH/ORP/reference, temperature, TGD, turbidity, chlorophyll, PAR. (From http:// www.hydrolab.com.)

5. Future Detection Technologies

The research and development of new generations of sensors, detection systems, and technologies is mainly focused on detection of environmental threats (chemicals, biologics, and radiological agents). Many of these technologies are specifically developed for detection of pollutants in air (samples). Only few of them are potentially appropriate for water contaminations applications. As for the present technologies, also in the future, the instruments of practical interest for water monitoring applications will be those which will operate unattended, for long periods as OCMS), at a reasonable price, so that large deployment of systems will be achievable. Examples of some promising development projects of future technologies are described below:

5.1. SANDIA WATER INITIATIVE

Sandia laboratories are developing new technologies for detection of chemical and biological contaminants in water. Two complementary technological approaches are the "combined preconcentrator and sensor for live waterborne pathogens", and the "real-time discriminatory sensors for water quality determination," initiated in October 2002.

5.1.1. Combined Preconcentrator and Sensor for Live Pathogens (http://www.sandia.gov/water/projects/preconcentrator.htm)

The "combine preconcentrator and sensor for live waterborne pathogens" system is based on application of an electric field across a two-dimensional array of insulated, microfabricated posts, to separate, isolate, and concentrate, microparticles (0.1–10 μm), based on their physical characteristics (polarization, geometry, and structure) (Fig. 6). The system is intended to operate continuously, at flows greater than 1 l/h, and provide concentrated samples to microfluidic separation channel. Coupling of the microfluidic separation channel to different sensor technologies such as the "MicroChemLab" (elaborated below) could provide in future basis for development of a new range of practical water monitors of water quality and detection of chemical and biological contaminants.

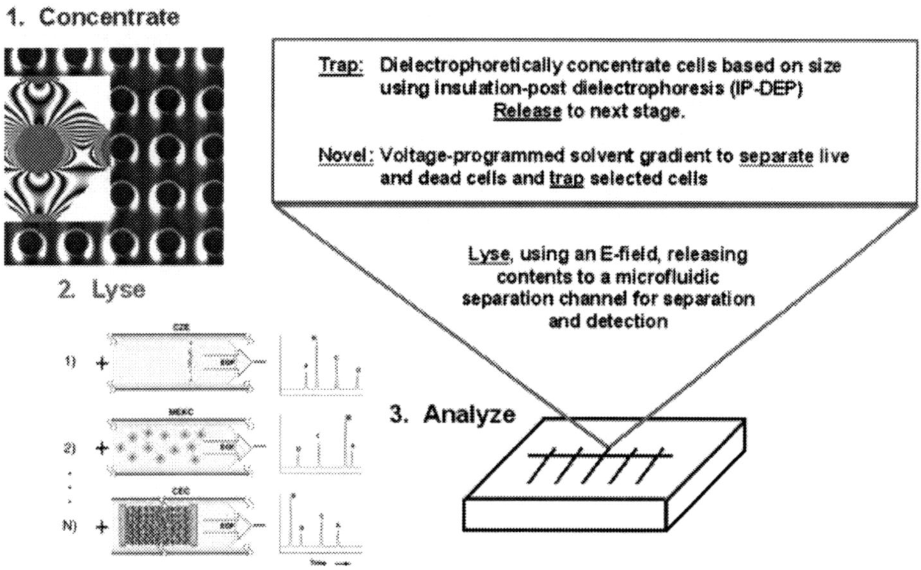

Figure 6. Principle of operation of the "combined preconcentrator and sensor for live waterborne pathogens," Sandia Laboratories

5.1.2. *Real-time discriminatory sensors for water quality determination: MicroChemLab (http://www.sandia.gov/water/projects/ChemLab.htm)*

This development project of Sandia is studying the possibility to implement the MicroChemLab technology for water quality monitoring, and real-time detection of contaminants (volatiles [VOCs] and semivolatile organics) via gas phase sampling (Fig. 7). This detection system, together with sampling components (e.g. the "combined preconcentrator," solid phase micro-extraction, or similar), could provide basis for cost-efficient monitoring systems of environmental and water contaminants.

The device demonstrated ability to analyze chemical weapon hydrolysis products in water.

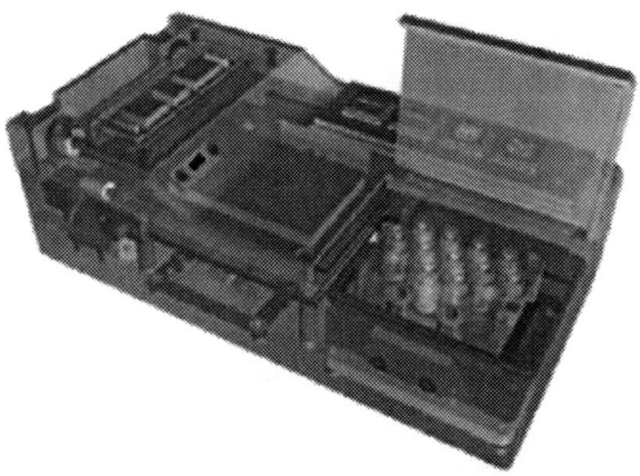

Figure 7. Sandia MicroChemLab system

5.1.3. *"EchoSensor" Semiselective Optical biosensors, Echo Technologies (Mass High Tech: The Journal of New England Technology, 2003)*

The Virginia-based company "Echo Technologies" is developing proprietary sensor technology for detection of biological and chemical agents in air and in water.

The company's core technology is a suite of optical biosensors that are capable of real-time detection and distinction between groups of biological agents such as bacteria, spores, and toxins in a few minutes.

The technology provides presumptive generic detection without use of added fluidics, and is designed to operate remotely with no user intervention.

In 2003, the company had a prototype product and targeted national security applications (military and civilian), including detection of environmental (air) and water contaminants.

6. Discussion

Due to the nature of the threat posed by terror contamination of water delivery systems and the short timetable (minutes to 1 h) in which effective steps can be taken to minimize the numbers of affected people and the damage to infrastructure, the prevailing approach is based on the implementation of surveillance measures and technologies for "risk reduction."

Such antiterror steps currently include improvement of physical security measures for critical assets (high-potential vulnerability to attacks), and should also include implementation of specific water QC technologies such as installation of OCMS with capabilities for detection (directly or through surrogate parameters) and real-time warning in presence of relevant contaminants.

Operation of those technologies should be part of the standard operating procedures for QC, and accompanied by upgrading and adaptation of the supervisory control and data acquisition (SCADA) systems for real-time monitoring of contaminants.

Despite the impressive technical progress in online water monitoring technologies and the promise of new instrumentation, detection with complete certainty of pollutants is expensive, and remains problematic. Nevertheless, meaningful reduction of risk is achievable.

References

1. ATSDR database on toxic chemicals, http://www.atsdr.cdc.gov
2. CDC list of biological contaminants, http://www.bt.cdc.gov/agent/agentlist.asp
3. Burrows, W.D. and Renner, S.E. (1999) Biological warfare agents as threats to potable water. Environmental Health Perspectives 107 (12):975–984.
4. Drinking Supply: Terrorists had eyes on water: Las Vegas Review-Journal 12 August 2004, http://www.reviewjournal.com/lvrj_home/2004/Aug-12-Thu-2004/news/24519286.html
5. EPA list of contaminants whose maximum allowable concentrations in water are regulated, http://www.epa.gov/safewater/mcl.html
6. Interim Voluntary Guidelines for Designing an Online Contaminant Monitoring System (2004) American Society of Civil Engineers (ASCE), December 9, 2004.
7. Jernigan, D.B., Raghunathan, et al. (2002) The National Anthrax Epidemiological Investigation Team. Investigation of bioterrorism-related anthrax, United States, 2001: *Epidemiological Findings*. Emerging Infectious Disease 8:1019–1028.
8. Mass High Tech: The Journal of New England Technology – 24 February 2003, Echo Technologies answers the military's call for biosensors, http://masshightech.bizjournals.com/masshightec/stories/2003/02/24/story9.html
9. Radiological contaminants, http://www.cfsan.fda.gov/~mow/intro.html
10. Rittman, B.E. and McCarty, P.L. (2001) Environmental Biotechnology, McGraw-Hill, New York.

4.2 THE ECOLOGICAL MONITORING OF POLLUTION AND MODELING OF HYDROECOSYSTEMS

TRAHEL VARDANIAN *
Yerevan State University, Yerevan, Armenia

Abstract: The paper examines the issue of the ecological monitoring of pollution and modeling of hydroecosystems. In the paper, an attempt was made to design a model for the ecosystem of a water body. To create the model we included all the environment-forming factors which directly or indirectly influence the ecosystem of a water body. The ecological factors are classified into three large groups: abiotic, biotic, and anthropogenic, each of them comprising numerous factors. The model we construct is a holistic system of differential equations, like a live organism – any change within the system will directly or indirectly affect the whole ecosystem. The equations represent one holistic model, which will function in the case of changing any parameter. This model also allows us to make forecasts based on different scenarios of ecological factors. It is applicable for analogical water bodies as well.

Keywords: water object, ecosystem, system of monitoring, ecological factors, modeling, live model

1. Introduction

In the concept of sustainable human development, water resources, and particularly their current ecological condition, have a significant strategic impact. To assess the current ecological condition properly, as well as to produce measures against its deterioration, it is necessary to establish a system of monitoring for the ecological condition of waters. The system must be based on the complex criteria, namely, quality and quantity of water resources, and flora and fauna. In order to obtain reliable information, a proper assessment of ecosystem, and to operate activities of the monitoring system well, it is important to make systematic observations and analyses on the regime and amount of river runoff, of water quality, and sources and substances polluting the water bodies.

2. Methods and Materials

To develop the program for ecological monitoring, it is necessary to coordinate the physical, chemical, and biological parameters, which aim at identifying and understanding the role and significance of each one for the improvement of the water ecosystem.

The circumstances of organizing abiotic and biotic observations at the same time and place are rather important. The latter will support the identification of their interrelation and will create a dynamic model within ecosystem.

For the ecosystem approach, long-term programs of monitoring are critical, since they will enable to reveal and evaluate the structure of cycles in ecosystems. It, certainly, is a long-term work and may take years of implementation and considerable expenditures.

In the Republic of Armenia, for instance, the ecological monitoring activities of surface waters, at the initial stage of their development, will be based only on the current network of observations, and later on they may extend simultaneously with the improvement of socio-economic situation in the country. It is worth noting that on the territory of Armenia (only 29,800 km^2), more than 200 observation posts and stations used to operate on rivers in different periods. Before the economic crisis in 1990, about half of the aforementioned number of posts and stations worked simultaneously. At the first sight it may seem that the obser-

Y. Sharan et al. (eds.), Water Supply in Emergency Situations, 101–104.

vation network is rather dense for this small area; however, when we take into account certain conditions, the picture is different.

Thus, considering the conditions of the complex mountainous relief of the territory (where the height varies from 375 to 4,090 m), the diverse geologic structure (volcanic and folded block-rocks), a relatively dense river network (k = 0.82 km/km^2), as well as the sheer number of rivers (the number of rivers and streams with length of 10 km or longer amounts to 380), the number of observation posts and stations is too small.

Considering the fact that in the last 10 years their number decreased by 30% and that chemical analysis of river water has hardly been done (except for single cases), we believe that it is difficult to establish a system of monitoring of the ecological condition of surface waters.

The establishment of such a system will be realistic only after a gradual rehabilitation of the economy of the country.

3. Obtained Results and Their Analysis

Having a monitoring system of the ecological condition of water objects, one can create a range of models.

Before the 1970s, numerous models were described around the world. But none of them had a clearly hydrological direction. The models of global problems, made by Tinbergen et al. (1976), also comprised natural components. Gabor's and Laslo's models had interrelation between nature and society in basis (Laslo, 1977; Gabor et al., 1981).

The above-mentioned works later served for building global models of ecological blocks.

Today, throughout the world there are a number of models encompassing many ecological blocks of natural objects and phenomena. However, when there is a need to describe a model of the ecosystem of a natural body, singling it out of the whole system, and simultaneously preserving the link, there are always a number of difficulties (Problems of Geography, 1986).

In describing a model of the ecosystem of a water body (a river or a lake), the following important conditions must be considered.

1. It is necessary to bind the model of the created ecosystem with other ecological blocks of the global model, which should have the required parameters of input and output.
2. The selected water body should have almost the same natural climatic conditions to operate the model.
3. The water body must be located within the physical geographical area, and not state borders of countries, in order to provide maximum precise forecasts.

Of course, the above-mentioned conditions are equally applicable for creating a model of ecosystem for any natural body.

To create the model we included all environment forming factors which directly or indirectly influence the ecosystem of the water body (henceforth to be called ecological factors).

All the ecological factors are classified into three large groups: abiotic, biotic, and anthropogenic, each of them comprised by numerous factors (Fig. 1).

To make the model function, it is necessary to have the mean annual indices of all ecological factors, which may be reflected integrally or in way of indices or several exponents. These exponents allow us to make differential equations between all the ecological factors and the water body (Fig. 2).

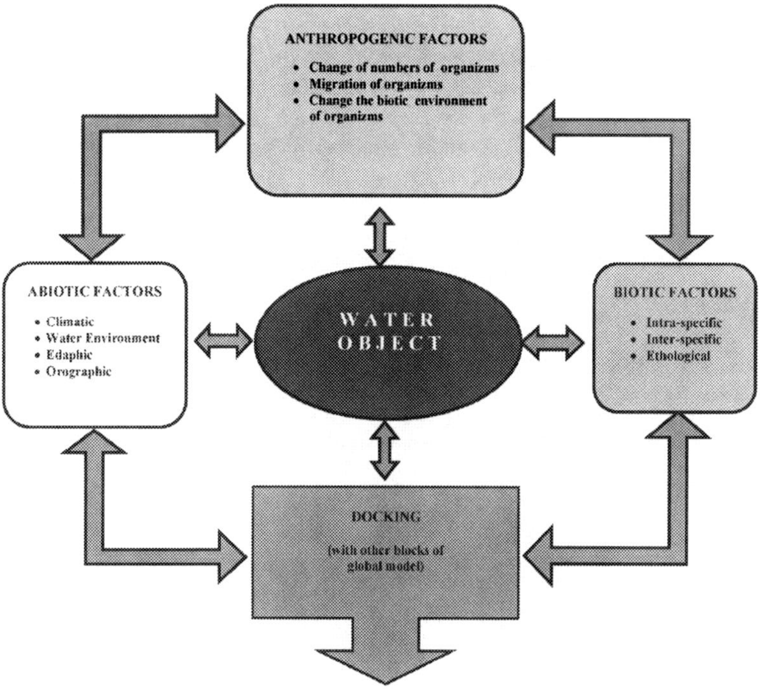

Figure 1. The chart of the model of the main ecological factors

Figure 2. "Live " model

4. Conclusions

Thus, our model is a holistic system of differential equations, like a live organism – any change within the system will directly or indirectly affect the whole ecosystem.

The equations represent one holistic model, which will continue functioning in case of changing any parameter.

This model also allows us to make forecasts based on different scenarios of ecological factors. It is applicable for analogical water bodies as well.

References

1. Gabor, D. et al. (1976) Olter l'eta della spreco, Milano.
2. Laslo, E. (1977) Goals for Mankind, Oxford University Press, New York.
3. The Problems of Geography (1986). Modeling of Geosystems, vol 127. Nauka, Moscow, 1986.
4. Tinbergen, Y. et al. (1976) Reshaping the International Order, Report to the Club of Rome, New York.

4.3 REAL-TIME ANALYSIS FOR EARLY WARNING SYSTEMS BASED ON MODELING CONCEPT

PETR INGEDULD,[1] EVŽEN ZEMAN,[1] PALLE LINDGAARD JORGENSEN,[2] AND MARSHA HORSNER[3]
[1]*DHI Hydroinform a.s., Prague 10, Na Vrsich 5, Czech Republic*
[2]*DHI Water and Environment, Agern Alle 11, DK-2970 Horsholm, Denmark*
[3]*DHI, Inc. 301 South State Street, Newtown, PA 18940, USA*

Abstract: Water treatment and distribution systems are highly vulnerable to degradation of quality and reliability of supply as a result of many factors: natural, accidental, and intentional. Among the potential intentional factors that the utility manager and operators have to plan for is the introduction of toxic contaminants into the water supply or disruption of water service through sabotage of key components of the infrastructure. Rapid recognition of the nature and location of such occurrences is vital to protect the integrity of the water supply and safeguard the consumers from potentially harmful contaminants, determine appropriate changes in supply and treatment strategy, and ensure compliance with environmental regulations. The utility manager and the operations staff must be given the proper tools as well as be trained to identify an event, locate the extent and potential danger to the public, and be prepared to react in a proper and timely fashion.

Rapidly developing sophisticated software and real-time instrumentation and monitoring systems provide the tools to design and develop early warning monitoring systems and to increase the preparedness of the water utility to react to such unexpected events. Proper integration of state of the industry hydraulic modeling systems, geographical information systems (GIS) for the water distribution network, and the installation of a supervisory control and data acquisition (SCADA) system for both water treatment plant and active element control, as well as the monitoring of critical points within the distribution system will be an invaluable resource for the operator to react to an event (real-time response) as well as to plan for possible future events (contingency planning).

Keywords: real-iime modeling, SCADA, online simulation, operations, hydraulic and water quality modeling, early warning systems, vulnerability of water supply, contingency planning, system integration

1. Early Warning Monitoring Systems

The goal of an early warning monitoring system is to reliably identify low probability or high impact contamination events (chemical, microbial, radioactive) in source water or distribution systems in time to allow an effective local response that reduces or avoids entirely the adverse impacts that may result from the client (Brosnan, 1999).

Requirements for the ideal early warning system:

- Provides warning in sufficient time for action.
- Cost is affordable.
- Requires low skill and training.
- Covers all potential threats.
- Is able to identify the source.
- Is sensitive to quality at regulatory levels.
- Gives minimal false positive or negative responses.

Y. Sharan et al. (eds.), Water Supply in Emergency Situations, 105–113.

- Is robust.
- Is reproducible and verifiable.
- Functions year-round.

A key component of early warning systems is the availability of a mathematical model for predicting the transport and fate of the spill or contaminant so that downstream utilities can be warned. However, a water quality model should be regarded as a guide only to what will happen, and increased monitoring during spill events is often critical to verifying the model and determining the fate of the spill and the safety of the supply.

The number of intentional threats and acts of sabotage against water supply systems is relatively small, and hoaxes are very likely. Intentional threats include:

- Destruction of a water supply system. Destruction of parts of the system can happen by physical destruction or computer hacking. Cyber attacks against various computerized components of water supply systems include attacks against SCADA systems.
- Contamination of the system with chemicals, microbes, toxins, or radioactive compounds. For example, nerve agents, cyanide, arsenic, and nicotine. The most likely points of attack for intentional contamination include posttreatment storage reservoirs, distribution tanks, and water mains (Haimes et al., 1998).

2. Real-Time Response

Simulation tools (i.e. well-calibrated hydraulic and water quality models) can be linked to SCADA real-time databases allowing for continuous, high-speed modeling of the pressure, flow, and water quality conditions throughout the water distribution network. Such models provide the operator with computed system status data within the distribution network. These "virtual sensors" complement the measured data. Anomalies between measured and modeled data are automatically observed, and computed values that exceed predetermined alarm thresholds are automatically flagged by the SCADA system. The operator, upon identification of an occurrence, can take appropriate action to either eliminate or contain the danger to public health or service interruption, or failing that, is able to map out the extent of the service disruption to guide both utility crews and emergency response units. Having taken corrective action, the operator can use the predictive modeling capability to extrapolate the future system performance. For example, in the case of introduction of contamination into the distribution system, the dispersion and dilution as a function of time can be calculated. By real-time monitoring, the operator can continually update and adjust the model. Similarly, alternate water supply strategies can be quickly modeled and evaluated, guiding the operator until the situation is remedied or stabilized.

3. Contingency Planning

The best time to prepare for an emergency, no matter how remote the possibility, is before it happens. The US Federal Government has required water suppliers to provide written emergency action plans. The utility engineer using the hydraulic modeling system has a powerful learning tool to assist in decision making and planning. Any number of scenarios may be mapped out and the appropriate responses documented for future use. Operators can be trained to use the model to simulate various scenarios in advance of a critical need, including failure of critical facilities and introduction of toxic substances into the distribution system. They can learn to evaluate and choose an appropriate strategy for operation of pumping stations or settings of the control valves and to determine the most appropriate response to unusual operating conditions, as well as being able to predict the potential for health hazard or service disruption. Such exercises can also be used to identify and quantify critical points in the system and provide input in the planning of capital improvements.

In summary, proper design and integration of hydraulic modeling software, GIS, and the SCADA system allows the water utility to plan for and react to scenarios to hopefully assure reliability of service and water quality in the face of whatever man or nature can throw at us, or, at the minimum, identify and contain the damage and disruption of service.

4. Hydraulic and Water Quality Analysis

MIKE NET is based upon the industry standard EPANET hydraulic and water quality algorithm. EPANET was developed by the Water Supply and Water Resources Division (formerly the Drinking Water Research Division) of the US Environmental Protection Agency's National Risk Management Research Laboratory. The hydraulic model used by EPANET is an extended period hydraulic simulator that solves the following set of equations for each storage nodes (tank or reservoir) in the system:

$$\partial y_s/\partial t = Q_s/S_v \quad (1)$$

$$Q_s = \Sigma_i Q_{is} - \Sigma_j Q_{sj} \quad (2)$$

$$h_s = E_s + y_s \quad (3)$$

along with the following equations for each link (between nodes i and j) and each node k:

$$h_i - h_j = f(Q_{ij}) \quad (4)$$

$$\Sigma_i Q_{ik} - \Sigma_j Q_{kj} - Q_k = 0 \quad (5)$$

Equation (1) expresses conservation of water volume at a storage node while Equations (2) and (5) do the same for pipe junctions. Equation (4) represents the energy loss or gain due to flow within a link. For known initial storage node levels at time zero, Equations (4) and (5) are solved for all flows q_{ij} and heads h_i using Equation (3) as a boundary condition. This step is called "hydraulically balancing" the network, and is accomplished by using an iterative technique to solve the nonlinear equations involved.

The method used by EPANET to solve this system of equations is known as the "gradient algorithm," Todini and Pilati [5], and has several attractive features. First, the system of linear equations to be solved at each iteration of the algorithm is sparse, symmetric, and positive-definite. This allows highly efficient sparse matrix techniques to be used for their solution, George-Liu [6]. Second, the method maintains flow continuity at all nodes after its first iteration. Third, it can readily handle pumps and valves without having to change the structure of the equation matrix when the status of these components changes.

After a network hydraulic solution is obtained, flow into (or out of) each storage node, q_s is found from Equation (2) and used in Equation (1) to find new storage node elevations after a time step dt. This process is then repeated for all subsequent time steps for the remainder of the simulation period.

MIKE NET's dynamic water quality simulator tracks the fate of a dissolved substance flowing through the network over time. It uses the flows from the hydraulic simulation to solve a conservation of mass equation for the substance within each link connecting nodes i and j:

$$\partial c_{ij}/\partial t = -(Q_{ij}/S_{ij})(\partial c_{ij}/\partial X_{ij}) + \theta(c_{ij}) \quad (6)$$

Equation (6) must be solved with a known initial condition at time zero and the following boundary condition at the beginning of the link, i.e. at node i where $x_{ij} = 0$:

$$c_{ij}(0,t) = \frac{\sum_{ki} Q_{ki} c_{ki}(L_{ki,} t) + M_i}{\sum_{ki} Q_{ki} + Q_{si}} \qquad (7)$$

The summations are made over all links *k,i* that have flow into the head node (*i*) of link *i ,j*, while L_{ki} is the length of link *k,i*, M_i is the substance mass introduced by any external source at node *i*, and Q_{si} is the source's flow rate. Observe that the boundary condition for link *i,j* depends on the end node concentrations of all links *k,i* that deliver flow to link *i,j*. Thus, Equations (6) and (7) form a coupled set of differential and algebraic equations over all links in the network. Water quality simulator uses a Lagrangian time-based approach to track the fate of discrete parcels of water as they move along pipes and mix together at junctions between fixed-length time steps. These water quality time steps are typically much shorter than the hydraulic time step (e.g. minutes rather than hours) to accommodate the short times of travel that can occur within pipes.

By employing these features, MIKE NET can study such water quality phenomena as:

- Blending water from different sources.
- Age of water throughout a system.
- Loss of chlorine residuals.
- Growth of disinfection by-products.
- Contaminant propagation events.

5. The Use of Models

Models of water supply networks (combined with GIS and SCADA) can be used as an instrument for increasing the public safety by providing answers to questions such as:

- How can we modify the water supply system or the operational procedures in order to reduce risks?
- How should we react, if an incident occurs?
- How do we get back to normal supply, when an incident has occurred?
- How will the supply of main be cut off if contaminated?
- What amount of time is required to flush each contaminated area?
- What neighborhoods are affected by each main?

6. How Can We Reduce Risks?

Any given tap receives water, which arrives though a number of pipes in the supply network, the transport route, and ultimately comes from a source ([2] Bunn, S., Helms, S., 2001). However, in order to achieve maximum supply security in case of pipe failures or unusual demand patterns (e.g. fire flows) water supply networks are generally designed as complicated, looped systems, where each tap typically can receive water from several sources and intermediate storage facilities. This means that the water from any given tap can arrive through several different routes and can be a mixture of water from several sources. The routes and sources for a given tap can vary over time, depending on the pattern of water use.

A model can show:

- Which sources (well-fields, reservoirs, and tanks) contribute to the supply of which parts of the city?
- Where does the water come from (percentage distribution) at any specific location in the system (any given tap or pipe)?

- How long has the water been traveling in the pipe system, before it reaches a specific location?

One way to reduce the risk – and simplify the response to incidents – is by compartmentalizing the water supply system. If each tap receives water from one and only one reservoir, pollution of one reservoir will affect one well-defined and relatively smaller part of the city. If a toxic substance is injected into any section of the water supply system, then one and only one part of the supply system will be polluted, thus reducing the potential risk in terms of the number of people involved.

Compartmentalizing the water supply system will reduce the spreading of toxic substances. On the flip side, it may increase the concentration of the toxic substance. It is also likely to have a negative impact on the supply of water for fire flow and on the robustness of the water supply network in case of failures of pipes or other elements. These problems can be eliminated, if the compartmentalization is done properly, allowing selected valves to be opened in case of fire emergencies or pipe failures ([4] Hosner, 2002).

7. How Should We React, If an Incident Occurs?

Cities are now (if not before) establishing emergency and preparedness plans covering this kind of incidents. US Federal Law now requires the preparation of written emergency action plans for water utilities serving 3,100 or more customers. A model is an invaluable tool in the preparation of such emergency plans. The model will be able to simulate a wide range of emergency scenarios, and the results can be condensed into very specific instructions for the emergency officers in charge at the time of the incident.

For instance:

A number of people in an area are reported ill with symptoms leading to suspicion that this might be caused by pollution of the drinking water. By looking up in an online GIS system containing preprocessed model results, the authorities have the following information readily available:

- What is the source(s) of the drinking water for the affected area?
- How long has the water been traveling from the source(s) to the taps?
- Exactly which part of the city is receiving water from the same source(s)?
- How has the toxic substance in all likelihood been spreading? Hence, who should be warned first, and where can we expect to find most of the casualties?

This information can of course lead to actions such as sealing off the affected area from not-yet affected areas, warning of people within the affected district, starting medical treatment of people living in the affected area, and setting up medical emergency centers.

Other precomputed information could include:

Assuming that several reservoirs were polluted simultaneously, where should we take samples in order to quickly discover pollution spreading from other reservoirs? If such samples are negative, then those segments were most likely not polluted and people living in those areas can be told that their drinking water is safe.

Or where do we sample within the polluted segment in order to find out exactly where the toxic substance was in fact injected? Based on concentration patterns, it is possible to rule out some locations and point towards the likely spots where the toxic substance can have been introduced into the water supply system.

8. How Do We Get Back to Normal Supply, When an Incident Has Occurred?

The term artificial recharge (AR) covers a range of technologies that typically utilize the natural cleaning capacity of natural subsoil systems to produce drinking water from surface water. The idea is to rapidly infiltrate surface water into the aquifer thereby increasing the groundwater formation and exploration possibilities. At many plants around Europe this is done in large plants where surface water is lead to large basins where it infiltrates. Traditionally, these plants are operated based on measurements of water quality on a regular basis and in many cases on real-time measurements of various flow-related parameters. This AR, or mixing together of polluted and fresh water to bring the polluted water to acceptable standards for use, is one method of dealing with an attack.

Cleaning the pipe system by flushing it is a relatively simple method, but the model is needed in order to ensure that pockets of polluted water will not remain in the system after the flushing. These reactions should be preplanned for each segment of the pipe system, leading to instructions such as:

- Flush for 3 h by this and this method. Then change the flow in this and this manner and flush for another 2 h.
- Turn off this main and drain via fire hydrants.
- To dilute, open this valve for water supply to this subdivision, etc.

Figure 1 shows the geographical extent of the pollution, at time 0, 2s, 4, and 24 h after release of the substance in the point marked with an arrow (SOURCE TRACING). Parts of the network are not exposed to the pollution at all, as these areas are supplied from other sources.

Figure 1. Spreading of pollution simulated by model

Figure 2. GIS layer used as a background and the model resulting file on the top of it with demand allocations in a readable form

A GIS used as a front end for the model results will enable the emergency response staff to point and click on a specific address and get information about sources, areas affected, etc. (Fig. 2).

Figure 3 shows another way of illustrating the spreading over time of a pollutant in a water supply system.

Figure 3. Spreading over time of a pollutant in a water supply system

9. Online Analysis of the Water Quality Parameters

Linking the hydraulic model to telemetry systems allowing the modeling of hydraulic, water quality, and energy parameters is of growing interest ([3] Cameron, R.W. et al., 1998). The DHI development team offers a new solution for water supply and distribution networks management. The goal of the project is to integrate the computational model MIKE NET with the central control system for solving problems related to water distribution network (Fig. 4). MIKE NET implements a mathematical model that captures the infrastructure and flow conditions in the network. This model is calibrated and validated in real time with data supplied by SCADA system. The integrated model monitors the pressure and flow conditions in the network in an automatic fashion – the energy parameters are evaluated in a parallel and

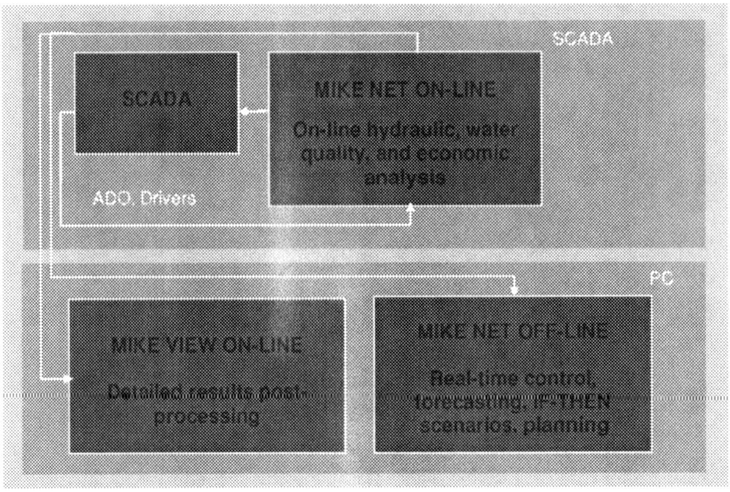

Figure 4. MIKE NET Online scheme

their optimization is enabled. Thus, any breakdown in the network can be readily identified, evaluated and rectified. The model also assists the practitioner to determine the most appropriate response to unusual operating conditions.

MIKE NET–SCADA provides the capability to model the water distribution system in real time, providing online modeling and monitoring of the system. This is essential when performing emergency response and it can greatly assist in confirming normal system performance, system trouble shooting, improvement of system operations, and projection of the current operating scenario. MIKE NET–SCADA can be linked to any existing SCADA monitoring system.

MIKE NET–SCADA consists of two modules:

- MIKE NET–SCADA online.
- MIKE NET–SCADA off-line.

The online module operates on top of the SCADA system and performs an online analysis of the system. The model results are stored back into the SCADA database; MIKE NET online viewer is used to display detailed model results.

The off-line module is used to model if-then scenarios, model system breakdowns, and predict system behavior based on the demand and control rules prediction. Microsoft access is the database used to store and maintain model alternatives.

10. Conclusions

Linking calibrated hydraulic and water quality model to SCADA real-time databases allows for continuous, high-speed modeling of the pressure, flow, and water quality conditions throughout the water distribution network. Such model provides the operator with computed system status data within the distribution network. These "virtual sensors" complement the measured data. Anomalies between measured and modeled data are automatically observed, and computed values that exceed predetermined alarm thresholds are automatically flagged by the SCADA system. The operator, upon identifying an occurrence, can take appropriate action to either eliminate or contain the danger to public health or service interruption, or failing that, is able to map out the extent of the service disruption to guide both utility crews and emergency response units. Having taken corrective action, the operator can use the predictive modeling capability to extrapolate the future system performance. For example, in the case of introduction of contamination into the distribution system, the dispersion and dilution as a function of time can be calculated. By real-time monitoring, the operator can continually update and adjust the model. Similarly, alternate water supply strategies can be quickly modeled and evaluated, guiding the operator until the situation is remedied or stabilizes.

References

1. Brosnan Thomas, M. (1999) Early warning monitoring to detect hazardous events in water supplies, An ILSI Risk Science Institute Workshop Report, ISBN: 1-57881-075-2.
2. Bunn, S. and Helms, S. (2001) Application of an expert system to control treated water distribution, Vodafone House, Auckland, New Zealand, sbunn@beca.co.nz, New Plymouth District Council, New Plymouth, helms@npdc.govt.nz.
3. Cameron, R.W., Barret, R.J., Cazottes, N, Jarrige, P.A, and Tocqueville, L. (1998) Link SCADA with network analysis system: what, why and how? Proceedings of the Third International Conference on Hydroinformatics, Copenhagen, Denmark, ISBN: 90 5410 983 1.
4. Hosner, M. (2002) Use of a Model to Help Secure a Water System, DHI North America Conference, Orlando, June.
5. Haimes, Y.Y., Matalas, N.C., Lambert, J.H., Jackson, B.A., and JRF Fellows (1998) Reducing vulnerability of water supply systems to attack. Journal of Infrastructure Systems 4 (4):164–167.
6. Ingeduld, P. and Turton, G. (2002) Online analysis of water supply and water distribution systems, NZWWA Conference, Auckland, New Zealand.
7. EPANET Methodology (1999) Water Supply and Water Resources Research, Division of the US Environmental Protection Agency's National Risk Management Research Laboratory, Cincinnati, OH.
8. MIKE NET User Guide (2001) DHI Water and Environment, Horsholm, Denmark.

4.4 APPLICATION OF REMOTE SENSING (OPTICAL AND SAR) TO MONITORING WATER RESOURCES

GARO MARDIROSSIAN
Space Research Institute at the Bulgarian Academy of Sciences, Bulgaria, Sofia 1000, 6, Moskovska St., P.O. Box 799, www.space.bas.bg; e-mail: mardirosian@space.bas.bg

Abstract: The presentation is dedicated to the potentials, implementation, and results from application of remote sensing techniques and equipment to monitor water resources for the purposes of hydrology as a whole.

Major attention is paid to optical studies and studies using synthetic aperture radar (SAR). The most suitable spectral ranges for solving various hydrologic problems are study of snow and ice covers, water basins and rivers, shallow waters and shelf, the volume and quality of water in artificial water catchments, and other examples, which are used to create models of water catchment surface change providing to calculate and monitor throughout the years the amount and quality of water contained therein. Subjects of such studies have been some dams and rivers of great importance to Bulgarian economy, such as the Studen Kladenets Dam, the Pchelina Dam, the Mesta River, and the Strouma River.

However, water, in all of its forms, causes as well great ecological catastrophes – floods, tsunami, and avalanches. This calls to study these extreme situations in all their aspects – forecasting, monitoring their development, evaluation of caused damages, etc. The potentials and prospects of remote sensing are illustrated, as an effective and operative means to combat water-caused catastrophes.

Attention is paid as well to ground-based (contact) information, required *a priori* in complex experiments for remote sensing of water catchments, without which the effective interpretation of remotely sensed data would be impossible.

In the end of the presentation, brief information is provided about the research and applied activity of the Department of Remote Sensing of the Earth at the Space Research Institute of the Bulgarian Academy of Sciences and their application in the studies of the geosphere – lithosphere, hydrosphere, and atmosphere during the last dozen of years.

Keywords: synthetic aperture radar, SAR, monitoring, remote sensing, water resources

1. Applicability of Remote Sensing in Hydrology and Glaciology

Major sources supplying water catchments:

- Rain and snow precipitation
- Snow and ice-melted water

Investigating the characteristics of ice and snow covers

An important snow-cover parameter is its water equivalent. Knowing it is very important in forecasting the water amount that would be obtained during ice melting. It is already well known that a clearly expressed relationship is available between water equivalent and the emission capacity of snow cover in the microwave region.

Y. Sharan et al. (eds.), Water Supply in Emergency Situations, 115–123.

The level of underground water

Determination of water supplies in atmosphere

The amount of rain that may pour over a given region may be assessed based on the amount of moisture contained in the atmosphere above this region. Here, infrared and microwave remote sensing images are used, whereas the microwave ones provide greater details.

Determination of water surface characteristics:

- Study of flooded areas
- Space glaciology

The typical properties of snow and ice and their high contrast provide for their effective study by remote sensing methods and techniques. The extensive introduction of such methods and techniques provided for the emergency of a new section of glaciology – *space glaciology*. Ice and snow characteristics are studied in the visible, infrared, and microwave range of the electromagnetic spectrum.

There are methods using space photos to calculate the melted-snow stock in river valleys based on the change in relative snow coverage of river basins and the configuration of seasonal snow boundaries. An important factor in snow-melting dynamics is air temperature above it and wind velocity. Operative information thereof is also obtained by Remote Sensing.

2. Spectral Ranges for Investigations

Table 1 shows the most suitable spectral ranges for remote sensing of water surfaces for the purposes of hydrology (according to various authors)

TABLE 1. Spectral ranges for remote sensing of water surfaces

Hydrological study	Wavelength (μm)
Snow cover in the plains	0.50–0.80
Snow cover in the mountains	0.80–1.30
Ice cover of rivers or lakes	0.44–0.52
	0.50–0.55
	0.79–0.84
	0.70–1.30
	8–14
River basins	0.68–0.76
	0.79–1.10
River floods	0.70–0.80
	0.80–1.30
Water surface	0.80–1.10
Water-flooded territories	0.70–0.80
	0.80–1.10
Coastline change	0.68–0.76
	0.78–1.30
Water layer of depth < 5 m	0.70–0.80
Water layer of depth < 15 m	0.60–0.70
Water layer of depth > 15 m	0.50–0.60
Shallow-water and shelf relief	0.52–0.55

3. Synthetic Aperture Radar

Synthetic aperture radar (SAR) is one of the most prospective instrumentation for remote sensing of the Earth, inclusive of water resource (Table 2). SAR provides visual information, regardless of the time of day or night, illumination, or meteorological conditions.

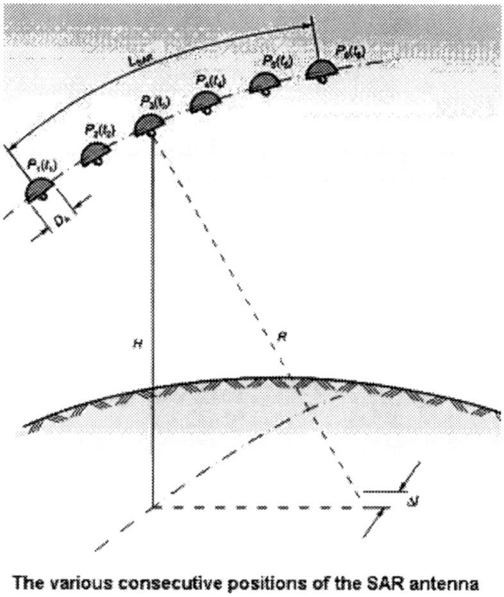

Figure 1. Major parameters of SAR system: H – height, D – diameter of antenna, R – distance from antenna to the measured element, L – length of synthesized aperture, Δl – space resolution

The elementary receiving–emitting antennae are located at several points of the orbit, i.e. they are synthesized as an antenna grid of very large (up to tens of kilometers) size (Fig. 1). At synphase summing up of the individual elements, the orientation diagram gets narrower.

Major SAR characteristics of the ERS-1 satellite (Fig. 2):

- Operates in the C-range (5.3 GHz).
- Spatial resolution of 30 m.
- Radiometric resolution of 2.5–16 dB.
- Data transfer rate of up to 105 Mbps.

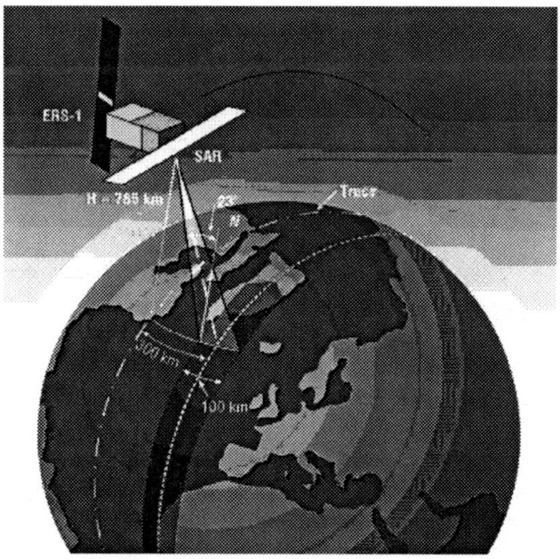

Figure 2. Major geometric parameters of SAR on-board of ERS-1 satellite

TABLE 2. SAR performances

Satellite	Launch years	Resolution (m)
SeaSat	1978	25
Cosmos 1870	1987	15–30
Almaz	1991	10–15
ERS-1	1991	26
ERS-2	1995	26
Envisat	2002	27
JERS-1	1992	25–30
Radarsat-1	1995	9

The National Aeronautics and Space Administration of USA (NASA) launched the recent LANDSAT satellite in 1993. The satellite can be found at a height of 705 km, repeating the image capture in 16 days above the same surface elements.

The collected data is transferred at the rate of 85 Mbps.

The geometric resolution of the image bands is 30×30 m^2, except for the thermal infra band (120 m) and pan band (15 m).

The sensor has seven spectral bands (it is called *multispectral scanner*) with the following spectrum intervals (Table 3):

TABLE 3. Spectral bands with spectral intervals

Image bands	Spectral range (μm)	Spectral response (color)
1	0.45–0.52	Blue-green
2	0.52–0.60	Green
3	0.63–0.69	Red
4	0.76–0.90	Near-infrared
5	1.55–1.75	Mid-infrared
6	10.40–12.50	Thermal infrared
7	2.08–2.35	Mid-infrared

4. Remote Sensing Monitoring of Water Quality

Images from the Advanced Spaceborne Thermal Emission and Reflection Radiometer (ASTER) aboard NASA's Terra satellite, launched in December 1999, illustrate the state of gradually decreasing water clarity in Lake Tahoe, one of the clearest lakes in the world.

In the image in Fig. 3a, acquired in November 2000, vegetation can be seen in red. The image in Fig. 3b acquired at the same time by a different spectral band of the instrument is color-coded to show the bottom of the lake around the shoreline. Where the data are black, the bottom cannot be seen.

Monitoring the lake's water clarity from boat measurements obtained since 1965 has discovered that the lake along the California–Nevada border has lost more than 1 ft of visibility each year. Details of the study can be found in the Lake Tahoe watershed assessment, a review of scientific information about the lake undertaken in February 2000. The most likely causes are increases in algal growth, sediment washed in from surrounding areas, and urban growth and development.

By combining historical and current ground-based measurements with space measurements from new instruments like ASTER, scientists are now continuously monitoring and gaining a better understanding of the circulation patterns and changes in Lake Tahoe's water clarity. Images like these from satellites, which are able to capture entire views of the lake and its 63

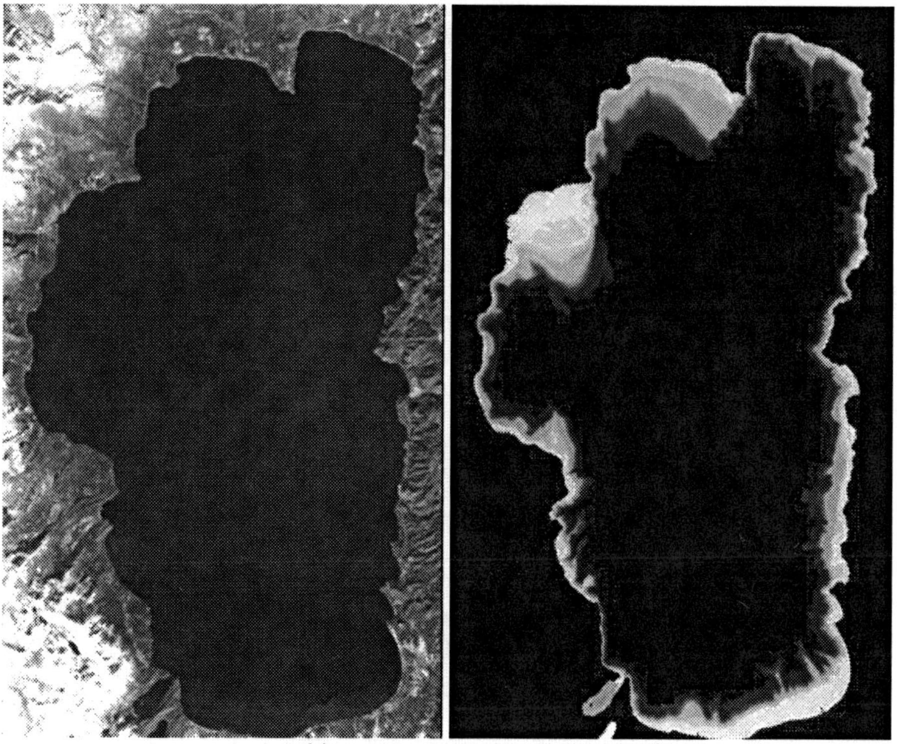

(a) (b)

Figure 3. Lake Tahoe - two different ASTER spectral bands acquired simultaneously

contributing streams, can be used to determine and monitor spatial variations in the lake's clarity over time. These images complement "point" measurements made by boats from one spot in the lake. Rafts and buoys verify the satellite images, and they help scientists develop and test circulation models.

5. Glacial lakes

According to a joint press release issued by NASA and the US Geological Survey, the great majority of the world's glaciers appear to be declining at rates equal to or greater than long-established trends. This image from the ASTER instrument aboard NASA's Terra satellite shows the termini of the glaciers in the Bhutan-Himalaya (Fig. 4). Glacial lakes have been rapidly forming on the surface of the debris-covered glaciers in this region during the last few decades.

It is believed that glaciers in the Himalaya are wasting at alarming and accelerating rates, as indicated by comparisons of satellite and historic data, and as shown by the widespread, rapid growth of lakes on the glacier surfaces. According to a 2001 report by the Intergovernmental Panel on Climate Change, scientists estimate that surface temperatures could rise by 1.4–5.8°C by the end of the century. A strong correlation between increasing temperatures and glacier retreat exists.

Figure 4. Space image of glacier

6. Space Research of *Iskar* Dam

The *Iskar* Dam (reservoir) is the major source of potable water for the City of Sofia, the capital city of Bulgaria, featuring over 1.2 million inhabitants (Fig. 5). The city spends annually about 220 million Cubic meter of water. The maximal volume of the dam is about 660 million cubic meter. The "dead" volume of the dam is 100 million cubic meter, at which water quality deteriorates abruptly.

Figure 5. Space image of Iskar Dam

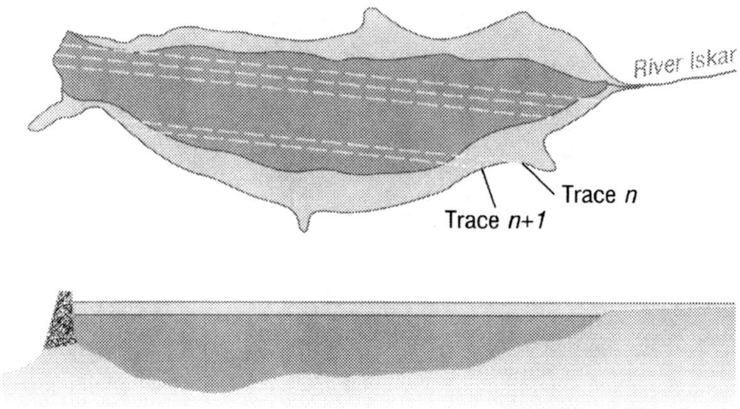

Figure 6. Boat mounted sonar profiling of the Iskar Dam reservoir and a resulting relief model of the dam's bottom (below)

During the "water crisis" of 1994, water volume reached up to 180 m^3, which called for imposing a severe water consumption regime in the capital city.

Remote sensing methods had been used for express evaluation of the amount of water in the *Iskar* Dam. The great change of the Dam's bottom relief as a result of washed in sediments and bottom landslides, called for its updating. This was achieved by a sonar mounted on-board a boat, which allowed to obtain individual profiles with boat trace step = 10 m (Fig. 6). Based on this data, a relief model of the dam's bottom was obtained. Based on the satellite-provided area of water surface "glass," express evaluation of the amount of water contained in the dam was made.

7. Space Research of Other Dams

The same aerospace methods are used in research for other Dams on the territory on Bulgaria. One of them is the *Malchika* Dam located in central north Bulgaria (Fig. 7) and Studen Kladenec reservoir (Fig. 8).

The image (Fig. 7) was obtained by "Landsat-5" in seven spectral channels, in the three basic colors R, G, and B (channels 4, 5, and 3). On it, the contours of various water depths are distinctly outlined. The darkest color (full absorption without reflection) corresponds to the greatest depth (above 4 m).

Figure 7. Contours of various water depths in Malchika Dam derived from a colour composite of three of the Landsat-5 spectral" channels (bands 4, 5, 3)

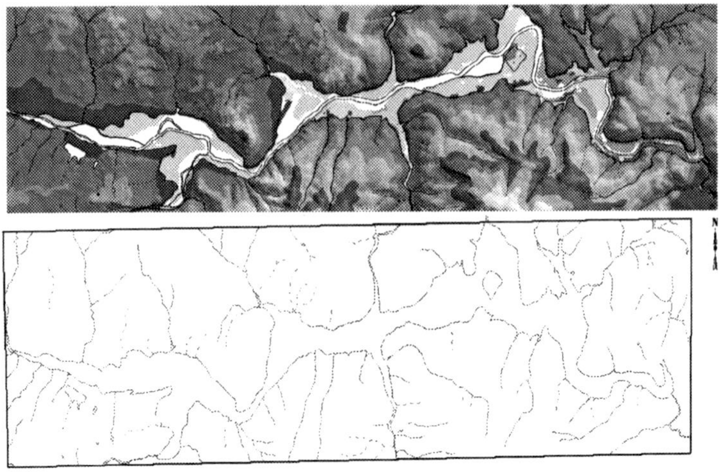

Figure 8. Studen Kladenec Reservoir

Monitoring water glass dynamics at various water levels and years:

Maximum level – 25.04 km^2
1977 – 12.88 km^2
1978 – 23.65 km^2
1992 – 18.80 km^2
2000 – 25.04 km^2

8. Water as Natural Hazards

However, water, in all of its forms, causes as well great ecological catastrophes – floods, tsunami, and avalanches. This calls to study all aspects of these extreme situations – forecasting, monitoring of their development, evaluation of caused damages, etc. The potentials and prospects of remote sensing are an effective and operative means to combat water-caused catastrophes (Fig. 9).

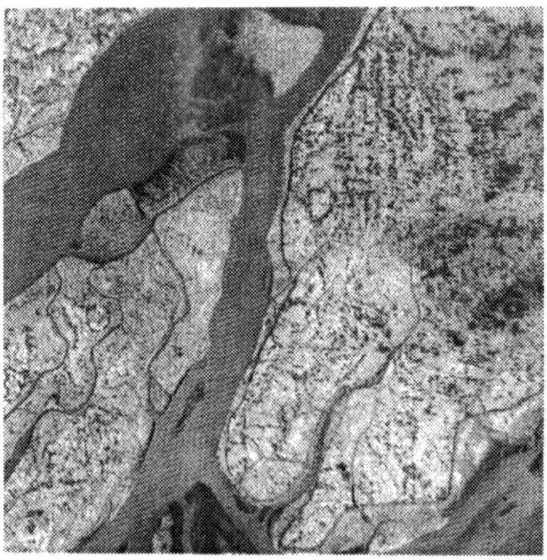

Figure 9. A satellite image of the delta of the Ganges River showing its discharge into the Bay of Bengal. The brown-red line along the coast is the dyke, protecting Bangladesh from floods.

9. Synchronous Aerospace Experiment for Hydrological Research

Ground-based (contact) information, required *a priori* in complex experiments for remote sensing of water catchments, without which the effective interpretation of remotely sensed data would be impossible (Fig. 10). The data obtained from ground-based studies may be used to solve various private tasks in the field of hydrology and to study water resources (Table 4).

TABLE 4. Hydrological investigation performed in synchronous aerospace experiment

Measured parameters	Units	Dynamic range	Absolute precision	Measurement interval (min)	Duration of the measurement (s)	Number of measurement points
Water level	cm		±1	60	10	1
Flow velocity	m/s	0 ÷ 5	±0.5	30	5	1
Water temperature	°C	0 ÷ 30	±0.1	15	10	1
Water salinity	‰	0 ÷ 25	±0.2	60		1
Water acidity	pH	3 ÷ 9	±0.5	30		1
Specific electric cond.	$\mu S/cm^2$	0 ÷ 4,000	±5	30		1

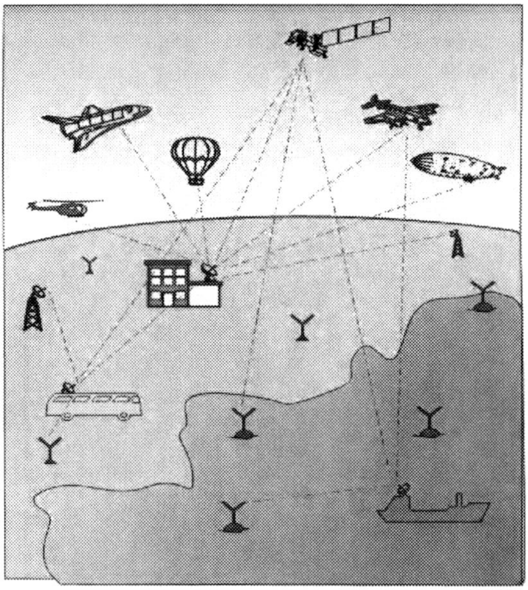

Figure 10. Illustration of a complex synchronous aerospace experiment for remote sensing of the Earth (maximal possible configuration)

References

1. Mardirossian, G. (2003) Aerospace Technique in Ecology and the Study of the Environment, Part 1, Marin Drinov Academic Publishing House, Sofia, 206.
2. Jelev, G. and Rumenina, E. (2001) Monitoring the Studen kladenets reservoir using air and space images Aerospace Research in Bulgaria, Space Research Institute, No. 16, pp. 97–106.

CHAPTER V

TREATMENT OF CONTAMINATED WATER

5.1 EXPERIENCE OF USING ENERGY-EFFECTIVE WATER DISINFECTION DEVICES

RASHID KHAYDAROV,[1] RENAT KHAYDAROV, AND BEKHZOD YULDASHEV
Institute of Nuclear Physics, Ulugbek, 702132, Tashkent, Uzbekistan
ROGER OLSEN
Camp Dresser and McKee Inc, 1331 17th Street, Suite 1200, Denver, CO 80202, USA

Abstract: The investigation described in this article examined an oligodynamic water disinfection method employing metal ions (Ag^+, Cu^{2+}, Au) generated by an electrolytic process. The efficacy of using different silver/copper/gold combinations (within the limits of current drinking water regulations) for killing *Escherichia coli*, *Legionella pneumophila*, *Salmonella*, *Vibrio cholerae*, and other pathogens were examined. The study investigated the dependence of bacteria killing time against metal ion concentration, different initial bacteria concentrations (from 10^3 to 10^{12} CFU/l), and the influence of different ion (Cl^-, SO_4^{2-}, S^{2-}, Fe^{2+}, Fe^{3+}) concentrations on the disinfection process. The disinfection devices with productivity of 5 and 50 m^3/h have energy consumption of 0.5 and 5 W-h, respectively. Results of tests of the method and devices conducted in different laboratories of the USA, Germany, Korea, Malaysia, Russia, and Uzbekistan, and experience of using the devices in these countries are described in the paper. The efficacy of a desalination device based on using the direct osmosis method has been examined. The device with productivity of 1 m^3/h consists of solar batteries with the capacity of 500 W, a water pretreatment unit on the base of fibroid sorbents, a water disinfection device with energy consumption of 0.1 W-h and a desalination device with power of 450 W. Thanks to the financial support of UNESCO. In March 2005, the pilot device with productivity of 1 m^3/h was installed in Turtkul village in the Aral Sea region to remove salts with total concentration of about 17 g/l.

Keywords: water supply, disinfection, oligodynamic method, pathogen, Aral Sea

1. Introduction

Water supply systems, water storage systems, cooling water in air-conditioners, swimming pools, etc. are one of the most attractive and vulnerable targets for bioterrorism. Furthermore, any natural cataclysms, for example, tsunami and earthquake, might cause destruction and bacterial pollution of water supply systems of cities and settlements. The fact that in these situations some settlements have not permanent access to electric energy makes the problem much more complicated. Different materials are used as the electrodes: silver [1][2], copper [3], and alloy of silver and copper in varying proportions ranging generally from 10/90 to 30/70 [3, 4]. The concentration of introduced metals in the disinfected water is of significant concern, particularly for potable applications. Metal limits of drinking water regulations vary in different countries. For example, copper is regulated by the US Primary Drinking Water Regulations at a concentration of 1.3 mg/l (action level, not maximum contaminant level); silver is regulated at a concentration of 0.1 mg/l as a secondary (i.e. nonenforceable) constituent because of its aesthetic effect, the discoloration of plumbing fixtures. The German and Russian drinking water regulations limit is 0.1 mg/l for Cu, and 0.01 mg/l, and 0.05 mg/l for Ag, respectively. Therefore the concentration of metal ions which is effective for disinfection sometimes is greater than the water regulation limits and consequently, the oligodynamic method cannot be used [5]. Moreover, concentrations and relations of metal ions in water (i.e. its disinfecting

[1] To whom correspondence should be addressed. Rashid Khaydarov, Department of Development of Devices, Institute of Nuclear Physics, Ulugbek, 702132, Tashkent, Republic of Uzbekistan; e-mail: green@sarkor.uz

Y. Sharan et al. (eds.), Water Supply in Emergency Situations, 127–132.

efficacy) strongly depend on factors such as the composition of electrodes, the water flow rate near the electrode surface, concentration of cations, and anions in the water.

Our investigations [6–8] have shown that the best disinfecting effect is obtained by using an alloy of silver–copper–gold composition with concentrations of metals in the ratio (70–90%):(10–30%):(0.1–0.2%), respectively. These empirically found out compositions allow providing considerable reduction of metal ions concentrations in the water in comparison with other combinations.

The objective of this work was to test water disinfection devices using our method [6–8] in laboratories and field conditions in different countries.

2. Materials and Methods

Experiments were conducted with the following bacterial cultures: *Escherichia coli* (272), (Strain 33218), (NCTC 10538), *Salmonella typhimurium* (251), fecal coliforms, typhoid – paratyphoid, *Legionella pneumophila*, *Salmonella*, and *Vibrio cholerae* in the temperature range of 18–36°C. The cultivation, culture enrichment, and the testing bacteria were performed following the standard methods for testing chemical procedures for disinfection.

Three types of water disinfections devices with the alloy composition of electrodes Ag/Cu/Au in the ratio 79.9%:20%:0.1% were made: a pen-size portable device using 6–12 V battery, the devices with productivity of 1 and 50 m^3/h using a solar battery, and 110–220 V AC, respect-tively.

The influence of Cl^-, SO_4^{2-}, S^{2-}, Fe^{2+}, Fe^{3+} ions on the efficacy of the disinfecting process was investigated at different concentrations. Stock solutions containing 1,000 mg of ions per liter were prepared by dissolving appropriate amounts of NaCl, Na_2SO_4, Na_2S, Fe $(NO_3)_2 \cdot 6H_2O$, $Fe(NO_3)_3 \cdot 9H_2O$ in deionized water and transferred to test solutions by using a dilution scheme. The water conductivity in the tests was not less than 10 μS/cm. In the experiments with low ion concentrations the conductivity was adjusted by adding a necessary amount of Na_2SO_4 stock solution.

A deluxe meter with ion measurements (Cole-Parmer Benchtop pH Meter) and Cole-Parmer combination ion-selective electrodes were used to determine concentrations of Ag and Cu ions in the water.

3. Results and Discussion

Disinfection can be described by the following empirical formula:

$$N(t) = N_0 \exp(-Ct/K)$$
$$\text{or } Ct = -K \ln(N/N_0)$$

where N_0 is the initial concentration of bacteria, N is the final concentration of bacteria, C is the effective concentration of metal ions in the water for a given ion collection, mg/l, K is the coefficient of resistance for a given ion collection, depending on bacteria types and the concentration of cations and anions in the water (that can react with disinfecting metal ions), and t is the time after introducing the metal ions into the water, i.e. disinfecting time (s). This formula shows that the concentration of metal ions C and time t necessary for water disinfection depend logarithmically on the initial concentration of bacteria N_0.

Table 1 summarizes coefficients *K* for different pathogens and disinfecting times. These data correspond to concentrations of silver, copper, and gold of 30 , 7.5, and 0.075 μg/l, respectively.

TABLE 1. Coefficient *K* and disinfecting time (initial concentration 1,000 CFU/l)

Pathogen	*K* (mg.s/l)	*t* (min)
Typhoid – paratyphoid	15–30	60
Legionella pneumophila		
Salmonella	8–15	30
Vibrio cholerae	8–15	30

The influence of different cations and anions (Cl^-, SO_4^{2-}, S^{2-}, Fe^{2+}, Fe^{3+}, etc.) in the water on the disinfecting process was investigated. Test results are given in Table 2. The initial concentration of *E. coli* in these experiments was 10^7 CFU/l, the concentration of Ag^+ ions in the water was 0.1 mg/l (the corresponding concentrations of Cu^{2+} and gold were 25 and 0.25 μg/l, respectively). These results show that Cl^- and Fe^{2+} ions have the most impact on the disinfecting process: they react with disinfecting metal ions with formation of AgCl and reduction of ions, decrease the concentration *C* of disinfecting metal ions and increase the disinfecting time *t*. The influence of other ions is not significant.

TABLE 2. Influence of cations and anions on disinfecting time of *Escherichia coli* (10^7 CFU/l)

Ions	Concentration of ions (mg/l)	Disinfecting time					
		15 min	30 min	60 min	90 min	2 h	24 h
Cl^-	2	300	3	<1	<1	<1	
	6	1,200	23	3	<1	<1	
	20	2,400	220	4	<1	<1	
	60	5,000	2,800	170	10	<1	
	350	10^4	5,000	890	70	12	<1
SO_4^{2-}	1.5		230	<1	<1		
	25		290	<1	<1		
	250		250	<1	<1		
S^{2-}	0.01	1,000	120	<1	<1		
	0.05	1,200	120	<1	<1		
	0.15	1,600	140	<1	<1		
	0.30	1,800	150	<1	<1		
Fe^{2+}	1.5	5,700	2,700	1,600			
	15	10^4	7,900	3,800			
Fe^{3+}	1.2	2,000	40	20	<1		
	11.12	2,200	110	9	<1		

The portable devices shown in Fig. 1 have been tested within the last 7-year period in different labs and universities. The study investigated the dependence of disinfecting time against effective concentration *C*, initial bacteria concentration N_0 (from 10^3 to 10^{12} CFU/l) and organism type. For example, in the Central Sanitary and Epidemiological Laboratory of the Ministry of Defense of Uzbekistan, this type of device was tested in disinfection of river water containing 10^8–10^9 CFU/l *E. coli*. The electrode insertion times *t* were 3, 5, and 10 min (the corresponding concentrations of Ag^+ and Cu^{2+} ions were 50 and 12.5 μg/l, 80 and 20 μg/l, 160 and 40 μg/l, respectively). The water samples were tested over a 20-day period following treatment by the device to determine the rate of bacterial regrowth. There was no detectable decrease of the disinfecting ion concentrations during 20 days. Some of the test results are given in Table 3.

Similar tests of the portable device have been made in the Department of Pathology of the Sains University of Malaysia in 1999, on pond water samples with fecal matters and bacteriological quality of 5×10^{12} CFU/l coliforms. Insertion time was 30 min and concentrations of

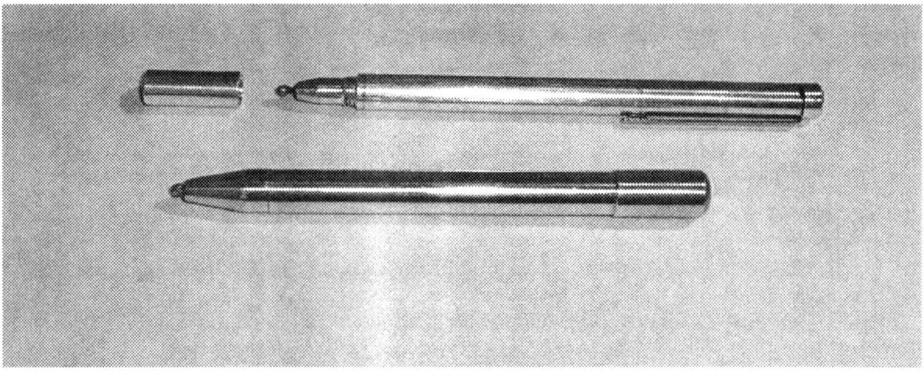

Figure 1. Portable disinfecting device

TABLE 3. Test results of the portable device

Disinfecting time	Initial concentration 10^9 CFU/l *E. coli*			Initial concentration 10^8 CFU/l *E. coli*		
	3 min insertion	5 min insertion	10 min insertion	3 min insertion	5 min insertion	10 min insertion
30 min	100–150	100–150	100–150	90–100	50	45
2 h	100–120	30	3	70	6	5
2 days	15	<1	<1	2	<1	<1
3–20 days	<1	<1	<1	<1	<1	<1

Ag^+ ,Cu^{2+}, and gold were 0.08 mg/l, 0.02 mg/l, and 0.8 μg/l. Samples were evaluated 1 h, 3 h, 1 day, and 1 week following treatment. "No growth" was reported for each of these samples, compared to mixed coliform growth in the control (i.e. non-disinfected) samples.

Tests of the portable device made in the Moscow Medical Academy (Russia), Envirotech Laboratories Inc. (Maryland, USA), Rogers Associates (Colorado, USA), EN Technology Inc. (Korea), in the Ministry of Public Health (Uzbekistan), etc., have confirmed the results described above.

The disinfecting devices with productivity of 1 m^3/h using solar battery were tested in disinfection of artesian well water containing 10^2–10^4 CFU/l *E. coli* by JDA International (Colorado, USA), by the Central Sanitary and Epidemiological Laboratory of the Ministry of Defense of Uzbekistan, by the Ministry of Public Health of Uzbekistan. Energy consumption of the devices was 0.2 W-h. The water samples were tested over a 30-day period following treatment by the device to determine the rate of bacterial re-growth. There was no detectable decrease of the disinfecting ion concentrations during 30 days. Some of the test results are given in Table 4.

250 devices were installed in manual artesian well water pumps in the Aral Sea region of Uzbekistan for disinfection of underground water (Fig. 2). Moreover, the devices were used in our water desalination minisystems installed in the Aral Sea basin. The productivity of the minisystem is 1 m^3/h and it works on solar batteries. The disinfection device in this minisystem is shown in Fig. 3.

TABLE 4. Test results of the device with the solar battery (final concentration of bacteria in CFU/l)

Time after treatment		
15 min	1 h	2 h
5.10^4	5	<1
9.10^2	<1	<1

Figure 2. Water treatment device installed into manual artesian water pumps in Aral Sea region

The disinfecting devices with high productivity of 5–50 m^3/h and with energy consumption of 0.5–5 W-h were tested in disinfection of artesian well water and swimming pool water containing 10^1– 10^3 CFU/l *E. coli* by JDA International (Colorado, USA), EN Technology Inc. (Korea), Institute of Medical Parasitology and Tropical Medicine (Russia, Moscow), State Center of Sanitary – Epidemiological Control in Moscow (Russia). The water samples

Figure 3. The disinfection device installed in water desalination systems with solar cells

were tested over a 60-day period following treatment by the device to determine the rate of bacterial regrowth. There was no growth of bacteria during 60 days. The device with productivity of 50 m^3/h installed in Korea is shown in Fig. 4. These devices have been used in USA at the Denver (Colorado) Zoological Park, in Malaysia after the tsunami for water disinfection of two fish farms and one water storage system and in Russia for treatment of swimming pools.

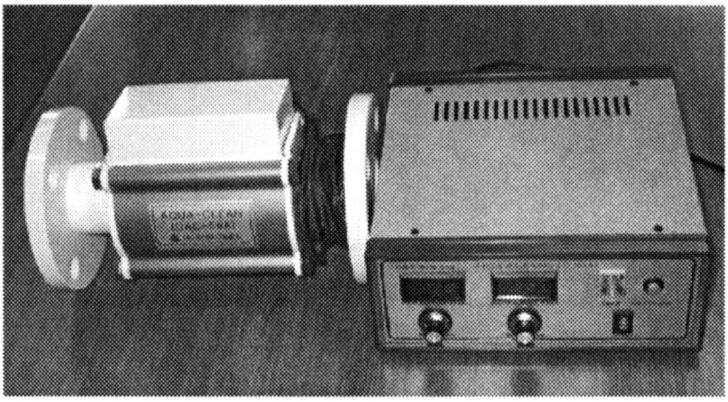

Figure 4. The disinfecting device with productivity of 50 m^3/h designed for South Korea

4. Conclusion

The results of these investigations demonstrate the efficacy of the oligodynamic method and devices for water disinfection. The pen-size devices with capability of 3 m^3/h of treated water can be used by the population to prevent bacterial contamination in emergency situations. The devices, with productivity of 5–50 m^3/h have small sizes, very low energy consumption of 0.5–5 W-h, work housing accumulators, solar batteries, or 110–220 V AC. They can be used for water disinfection of potable water systems, water storage systems, cooling water in air-conditioners, swimming pools, etc. to prevent consequences of bioterrorism and any natural cataclysms.

References

1. Kulsky, L.A. (1982) The Silver Water, Kyiv, Naukova Dumka, Ukraine.
2. Silver, S., Phung, L. (1996) Bacterial heavy metal resistance: new surprises. Annual Review of Microbiology 50:753–789.
3. Munson, G.P. et al. (2000) Identification a copper-responsive two-component system on the chromosome of *E. coli* K-12. Journal of Bacteriology 182:20.
4. The Silver Institute (1999) Silver: The Healthful Metal, Washington DC, December 31.
5. Rohr, U. et al., Impact of silver and copper on the survival of amoeba and ciliated protozoa in vitro. International Journal of Hygiene and Environmental Health 203 (1):87–89.
6. Khaydarov, R.A. and Malyshev, S.B. (2001) Patent of Russia, Water Disinfecting Device, N 2163571, February 27, 2001.
7. Khaydarov, R.A, Yuldashev, B., Korovin, S. and Iskandarova, Sh. (1997) Patent of Republic of Uzbekistan, Water disinfectant device, N5031, September 8, 1997.
8. Khaydarov, R.A., Khaydarov, R.R., Olsen, R.L. and Rogers, S.E. (2004) Use of electrolytically generated silver, copper and gold for water disinfection. Journal of Water Supply RT-Aqua 53:567–572.

5.2 USE OF NATURAL MATERIALS FOR MUNICIPAL WASTEWATER TREATMENT

GAGIK TOROSYAN,* ALEKSANDER ISAKOV, AIDA ALEKSANYAN, AND MERI ZAKARYAN
State Engineering University of Armenia 105 Teryan, 375009 Yerevan, Republic of Armenia

Abstract: This paper presents the results of research in the field of the use of natural materials – zeolites (mordenite, clinoptilolite, or zeolite containing tufas) and agricultural by-products (the by-products include soft lignocellulosics such as peach seeds-based activated carbon) as sorbents and as promising materials for raising the quality of water. The advantages of zeolites and agricultural by-products in comparison with other sorbents are their low cost, their availability, and others.

Keywords: natural materials, zeolite, agricultural by-product, activated carbon, peach seeds-based activated carbon, adsorption, adsorbent, wastewater treatment, organic solution, natural zeolite, mordenite, clinoptilolite, ammonia, organic impurities, phenol, aniline, oil products

1. Introduction

Municipal wastewater is characterized by physical characteristics as well as organic and inorganic contaminants. Physical characteristics include color, odor, and turbidity caused by dissolved or suspended solids. Chemically, wastewater is composed of organic and inorganic compounds as well as various gases. Organic contaminants include dissolved or undissolved volatile organic compounds (VOCs) and nonvolatile compounds – carbohydrates, proteins, fats, greases, surfactants, oils, pesticides, phenols, etc. Inorganic pollutants include compounds of trace minerals, sulfides, chlorides, nitrogen, and phosphorous.

Municipal wastewater must undergo primary, secondary, and tertiary treatments before being discharged, whether it has been used for domestic or industrial purposes. Following the secondary treatment, wastewater undergoes a tertiary treatment, which may be physical, chemical, biological, or a combination of two or more of these preceding methods. However, in most cases physical methods alone have been found sufficient to successfully complete tertiary treatment. Physical treatments include membrane separation, chemical coagulation, and adsorption by adsorbents, which have found wide applications in wastewater treatment.

In the present paper, we introduce our results from research in the field of water treatment process from difference type of pollutants, by using as physical treatment method adsorption on surface.

2. The Use of Zeolites for Water Treatment

The increasing demand for ion exchange materials as ecological problems stimulate the intensive study of natural and modified zeolites since they are considered to be cheep modified sorbents [1–4]. Natural zeolites have been utilized as chemical sieves, gas absorbers, odor control and filters in water, and in wastewater treatment processes.

The technological stability of zeolites as sorbents is determined first of all by characteristics such as mineral and chemical composition, sorption ability and then mechanical, physical characteristics, and subsequently their filtering properties [1].

* To whom correspondence should be addressed. Gagik Torosyan, Chemical Technology and Environmental Engineering, Department of State Engineering University of Armenia, 105 Teryan, 375 009 Yerevan, Republic of Armenia, e-mail: gtorosyan@seua.am

Y. Sharan et al. (eds.), Water Supply in Emergency Situations, 133–138.

Natural zeolites can be used in the following fields of water treatment [2]:

- Removal and recovery of NH_4.
- Removal and storage of radionuclides.
- Removal and storage of heavy metals.
- Removal of organic compounds.

The advantages of zeolites use in Armenia, in comparison with other sorbents are their great reserves and a unique complex of technological properties which allows us to modify their natural origin in various directions, including regeneration and utilization [5]. The application of the Armenian natural zeolites in the processes of water preparation has been scientifically approved according to the all-round evaluation of mechanical, physical, physical–chemical, and technological properties of the produced zeolites. It has been established that the natural zeolites and their combination with other systems is capable of removing metal ions from water. The consistency in the extraction of ions of iron, magnesium, calcium, zinc, copper, nickel, cobalt, lead, and ammonium from water has been established. The efficiency of the sorption mechanism of components from water, the filtering parameters, the length of contact liquid and solid phases ratio, and other factors are obtained.

Chemical stability and mechanical strength of natural Armenian zeolites – mordenite and clinopilolite meet the requirements of filtering materials.

In the present paper, ammonia, phenol, and aniline sorption from water solution have been investigated. Phenols and aromatics (benzene, toluene, xylenes, and others) when discharged in open reservoirs destroy the microflora and have negative impacts on human health. The principal method to diminish the presence of phenols dissolved in the water is intensive purification and reuse.

3. Using Zeolites in Present Research

Zeolites occur in nature in specific kinds of rocks. Zeolite-rich rocks are widespread in northern part of Armenia, occurring in much extended geological formations. The zeolite types are exclusively clinopilolite – in Idjevan (northern-east of Armenia) and mordenite in Shirak (northern-west of Armenia). Taking advantage of their high zeolite content, high cation exchange capacity, and selectivity, many agricultural applications of Armenian zeolites have been proposed recently. Some of Armenian natural zeolites were categorized from the point of view of chemical composition, type of structure and chemical, thermal and radiation resistance. These natural zeolites produced in Armenia were employed for this study.

4. Using Agricultural By-Products in Present Research

The abundance and availability of agricultural by-products make them good candidates as precursors for activated carbons. It has been shown that granular activated carbons based on agricultural by-products derived from pecan shells and sugarcane bagasse, have the potential for use in sugar refinery [6, 7].

There is limited literature on the use of agricultural by-product-based activated carbon for wastewater treatment processes.

In the present study we use peach seeds-based activated carbon.

5. Water Treatment for Ammonia Removal by Zeolites

The ammonium in water and wastewater can be toxic to the aquatic life and should be removed to prevent environmental pollution. Several processes are currently used for the removal of ammonium from aqueous solutions. Ion exchange with natural zeolites is among the methods most commonly employed.

Zeolites are materials well known for their ability to remove ammonia preferably from wastewater [8–12]. One of the aims of this work is to investigate ion exchange of ammonium $(NH_4)^+$ with zeolites obtained from the above-mentioned deposits in Armenia. The ion exchange capacity of natural zeolites, especially from Armenian deposits, offers the possibility for their application to the purification of ammonia-contaminated water.

Table 1 presents the best results for ammonia adsorption from water solution on zeolites which we used.

TABLE 1. The adsorption of NH_3 (g) on the zeolites/10 g

No.	Zeolite	1 day	3 days	5 days	7 days	10 days
1	Mordenite	10.38	10.51	10.62	10.70	10.76
2	Mordenite[a]	10.86	11.02	11.15	11.22	11.38
3	Mordenite[b]	11.14	11.71	12.13	12.46	12.98

[a]Treated by 0.5 N $CaCl_2$
[b]Treated by 1 N $CaCl_2$

6. Wastewater Treatment for Organic Impurities Removal with Zeolites and Agricultural By-Products

In this section, the results of the research on the application of zeolites as sorbents for phenols and aniline in water [13] are presented. The type of zeolites used were mordenite from Shirak region in Armenia, the H-form of this mordenite, zeolites ZSM-5 and silica KSK, and also agricultural by-product-based activated carbon. The linear dependence between the concentration of phenol in a water solution and the appropriate molar refraction is preset at 20°C (phenol concentration in water is 0.05 M). It was established earlier, that the sorption in these conditions increases and has a linear dependence on molar refraction. From the graphic dependence the amount of adsorbed phenol is determined. The phenol content in the water before and after the sorption was corrected with liquid chromatography.

The results are given in the Table 2. It is necessary to note that partial sorption of water /1–2 ml from 10 ml of a solution for 4 h sorption of a solution on sorbents [14] takes place.

As it becomes obvious from Table 2 the most active sorbents appeared to be is H-mordenite, synthesized zeolite – ZSM-5 and cationit. The silica KSK is an active phenol adsorbent from CCI4 solution, but in this case has appeared inefficient owing to absorption of plenty of water. Usual mordenite is also inefficient in the given process. The H-form of mordenite and cationit show activity, where, on all probability, the formation of hydrogen bonds takes place.

TABLE 2. The sorption of phenol from a water (0.05 M)* solution on sorbents/temperature 20°C, duration 4 h

No.	Sorbent	N_D^{20} after sorption	sorption (g phenol/g sorbent)
1	KSK	1.3320	0.0236
2	Mordenite	1.3320	0.0240
3	H-Mordenite	Full sorption	0.0380
4	ZSM-5	Full sorption	0.0375
5	Granular activated carbon**	1.3315	0.0289
6	Cationit	Full sorption	0.0380

*N_D^{20} for initial phenol solution is 1.3324
**Peach seeds-based activated carbon

The linear dependence between the concentration of aniline in a water solution and the appropriate molar refraction also exists. The measurements were carried out in a concentration 0.01 M in water. The aniline content in water before and after the sorption was also corrected by liquid chromatography and ultraviolet (UV) spectrum.

The results are given in the Table 3 and Fig. 1.

H-clinoptilolite and cationit are active. The formation of hydrogen bonds takes place, probably, here also.

TABLE 3. The sorption of aniline from a water solution on sorbents/concentration 0.01 M, temperature 20°C, duration 4 hours/*

No.	Natural supporters	N_D^{20} after sorption	Aniline sorption(%)
1	H-mordenite	1.3291	75
2	H-clinoptilolite	1.3289	100
3	Granular activated carbon**	1.3290	98
4	Cationit	1.3289	100

*N_D^{20} for initial aniline solution is 1.3300
**Peach seeds-based activated carbon

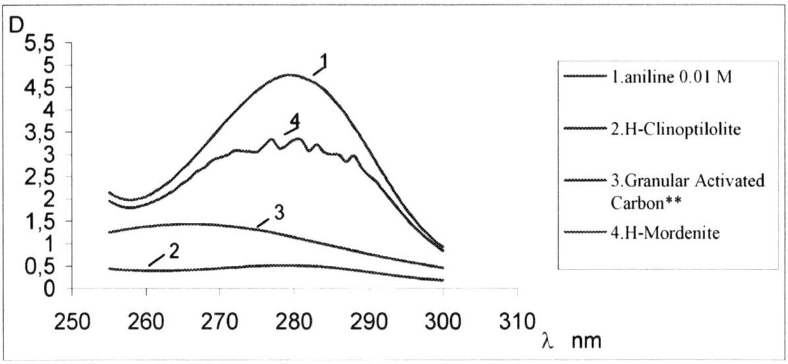

Figure 1. Aniline adsorption on the zeolites and AC

7. Treatment of Water-Soluble Oil Products Using Different Samples of Zeolites

In experiments, we compared the sorption of water-soluble oil products by different samples of zeolites – zeolites tuff, treating zeolites, and modified zeolites. Here, we used clinoptilolite containing tuff from the Idjevan deposits. They are modified by water-soluble polymer as polyethleneimine (PEI). The oil products' content in the water before and after the sorption was estimated by gravimetric method, after making a correction by gas–liquid chromatography. The results are presented in Table 4.

The oil products' content in the water before and after the sorption was estimated by gravimetric method, after making a correction by liquid chromatography.

TABLE 4. Distribution coefficient of the oil products K_a (mg/g) during the sorption on clinoptilolite and others

Sorbent	K_a/light organic[a]	K_a/heavy organic
Clinoptilolite-tuff	50	300
Clinoptilolite-treating	160	540
Clinoptilolite-modified	350	680

[a]Conditions t = 2 days, concentration of oil product in water – 70 mg/l

TABLE 5. Sorption of the oil from water solutions on sorbents

Sorbent	Weight of sorpted oil product (mg)[a]	Percentage of sorpted oil product
Clinoptilolite-tuff	20	30
Clinoptilolite-treating	32	45
Clinoptilolite-modified	50	70

[a]Oil product concentration in water 70 mg/l, 100 g sorbent, V water – 1,000 ml

8. Experimental Part

8.1. WATER TREATMENT FOR AMMONIA REMOVAL USING ZEOLITES

In this study, clinoptilolite from Idjevan region were prepared by heated pile method, using 0.5 and 1 N solution of $CaCl_2$. The natural zeolite – clinoptilolite and its cationic form were dehydrated at 110°C. After sorption of ammonia vapor produced by 1 N NH_3 solution by natural zeolites and its treated forms for 10 days at room temperature, the adsorption data were plotted.

The ammonia adsorption by zeolites was investigated by infrared (IR) spectroscopy. Mordenite from the Shirak region have not the same activity for ammonia adsorption.

8.2. WASTEWATER TREATMENT FOR ORGANIC IMPURITIES REMOVAL USING ZEOLITES AND AGRICULTURAL BY-PRODUCTS

For the liquid chromatography a higher-effective liquid chromatography (HELCh), detector Waters 486, controller Warers 600S, Pump, Waters 626, colon 250 × 4 mm, Si-100 C 18, P 150 Bar, V 1 ml/m, mobile phase acetonitril to water (50:50), detector UV-254) were used.

UV spectrometry was conducted on "Specord " UV spectrometer.

8.2.a. One gram of sorbent was added in 10 ml solutions of phenol in water. The mix was carefully shaken for a period of 4 h. The measurements of the molar refraction of a solution were carried out before and after sorption. The between primary and final amount of absorbed phenol is found by measuring the difference of concentrations of the organic solution.

The same measurements were carried out for aniline in concentration in 0.01 M. In this case, the results have been checked by liquid chromatography and UV spectrum dates also.

8.2.b. One gram of sorbent was added on a 10 ml (0.01 M) solution of aniline in water. The mix was carefully shaken for a period of 4 h. The measurements of molar refraction of a solution were carried out before and after sorption between primary and final amount of absorbed phenol is found by measuring the difference of concentrations of the organic solution.

9. Conclusion

The research in ammonia and organic pollutants sorption by Armenian natural zeolites and agricultural by-products-based granular activated carbon has been found to have advantages over other methods.

A convenient method for successful sorption of phenols and aniline from water has been offered. The given method can be applied at rather low initial concentration for other aromatics (aromatic compounds benzene, toluene, xylenes, ethylbenzene [BTEX]).

References

1. Donald W.B. (1974) Zeolite Molecular Sieves: Structure, Chemistry and Use, Wiley-Interscience/ Wiley, New York/London/Sydney/Toronto, p. 782.
2. Collela (1999) Porous materials in environmentally friendly processes, Studies in Surface Science and Catalysis, vol 125. Elsevier, Amsterdam, pp. 641–655.
3. Galarneau, A, Di Renzo, F., Fajula, F., Verdine, J. (eds) (2001) Zeolites and mesoporous materials at the dawn of the 21st century, Proceedings of the 13th International Zeolite Conference, Montpellier, France, July 8–13, Series Studies in Surface Science and Catalysis, Advisory editors: Delmon, B. and Jates, J.T., vol 135 and Recent Research Reports for 13th IZC.
4. Misaelidis, P. (ed.) (2002) Zeolite 02, 6th International Conference on the Occurrence, Properties and Utilization of Natural Zeolites, Book of abstracts, Thessaloniki.
5. Torosyan, G., Hovhannisyan, D., Shahinyan, S., Sargsyan, H. (2002) Armenian zeolites and its possibilities in industrial and municipal waste-water cleaning. Ecological Journal of Armenia 1 (1):93–96.
6. Ahmedna, M., Johns, M., Clarke, S.J., Marshall, W.E. and Rao, R.M. (1997). Journal of the Science of Food and Agriculture 75:117–124.
7. Ahmedna, M., Marshall, W.E., and Rao, R.M. (2000). Bioresource Technology 71 (2) 113–123.
8. Dyer, A. (1988) An introduction to Zeolite Molecular Sieves. Wiley, Chichester, p. 83.
9. Liberti, L., Lopez, A., Amicarelli, V. and Boghetich, G. (1995). In: Ming, D.W. and Mumpton, F.A. (eds) Natural Zeolites 93: Occurrence, Properties, Use, International Committee on Natural Zeolites, Brockport, New York, p. 351.
10. Kallo, D. (1995) International Committee on Natural Zeolites, Brockport, New York, p. 341.
11. Piper, R.G. and Smith, C.E. (1984) in: Pond, W. G. and Mumpton, F.A. (eds) Zeo-Agriculture: Use of Natural Zeolites in Agriculture and Aquaculture, Westview Press, Boulder, CO, p. 223.
12. Lahav, O. and Green, M. (1998) Ammonia removal using ion exchange and biological regeneration. Water Research 32 (7):2019–2028.
13. Water: Ecology and Technology (2000) 4th International Congress, ECWATECH-2000, Abstracts, Moscow, 30 May–2 June, 2000.
14. Sargsyan, S., Grigoryan, A, Harutjunyan, S., Torosyan, G. (2000) The phenol sorption on the zeolites, Bulletin of Armenian Constructors 2 (18):30–32.

CHAPTER VI

LEGAL ASPECTS

6.1 SUPPLY OF WATER TO THE POPULATION OF AZERBAIJAN IN EMERGENCY SITUATIONS AND ITS LEGAL AND INSTITUTIONAL ASPECTS

RAFIG VERDIYEV[1]
Scientific Research Institute on Hydrometeorology of Ministry of Ecology and Natural Resources, Baku, Azerbaijan, 36

Abstract: This article is directed at providing some recommendations for improvement of water supply for the population of Azerbaijan in emergency situations. This is possible through application of advanced water management methods and water use technologies based on the experience of leading countries in the world. In order to support implementation of a national program on improvement of the water sector, an action plan was developed. This plan reflects problems existing in water provision to the population and was approved by the government of the Azerbaijan Republic. Parallel with that, comprehensive environmental legal and normative acts have been adopted during the period of independence. Although they reflect the important economic and political changes, there are many difficulties with their implementation, connected mainly with absence of adequate water management policy, resources, and technology. Therefore, this article indicates that there is need for support of donor countries and international financial institutions for rehabilitation of water supply and wastewater treatment system of different settlements of the country.

Today, there is entire utilization of the water system in Azerbaijan as is the case with other countries of the South Caucasus, created during the Soviet times. Water provision to economy sectors and population is carried out by means of active old and inadequate hydrotechnical constructions. Often many water intake facilities and treatment systems are inoperable during high flooding and water turbidity, and extreme pollution. For example, on the September 8, 1999, turbidity at the mouth of the Kura River, where the water intake facility for provision of Baku city is constructed, had increased tens of times as compared to previous days, reaching 9,817 g/3 m^3. As a result, the water treatment system could not operate in its normal regime causing interruptions in the water supply of Baku and Sumgait cities and degradation of the quality of water. There are also many other such cases, the causes are entrance of highly polluted waters from urban areas, industry, and agricultural lands.

The main directions for effective water provision to the population of Azerbaijan in emergency situations in the context of sustainable development should include implementation of national and regional programs on water resources management as given in this article. An important step for improvement of water management in emergency situations is the creation and realization of a strong legal and institutional basis that would provide for attaining the required quality and quantity of drinking water, development of projects on determination, and use of alternative water sources in the case of emergency situations and use of modern technologies and methods of water provision. Some options for improved water supply in the case of emergency are described for different regions of the country.

Keywords: water management, flooding, integrated water resources management, emergency water supply, early warning system, water legislation

[1] To whom correspondence should be addressed. Rafig Verdiyev, SRIH, 36, Huseynbala Aliyev str. Apt. 52, 3rd microdistrict, Baku, Azerbaijan; e-mail: rafig2000@mail.ru

Y. Sharan et al. (eds.), Water Supply in Emergency Situations, 141–157.

1. Introduction

This article is directed at providing some recommendation for improvement of water supply for the population of Azerbaijan in emergency situations through, capacity building, technology transfer, application of advanced water management methods, and by creating legal and institutional frameworks for the development of Integrated Water Resources Management Plans at the national level and for the whole Kura–Araz River basin.

After the collapse of the Soviet Union, the countries of the South Caucasus gained their independence. However, they faced problems associated with national and transboundary water management. Transboundary water management remains one of the key issues leading to conflict in the region today. The scarcity of water especially in downstream areas is a major problem from both a socioeconomic and environmental point of view and often creates emergency situations, which result in provision of water only during for a few hours a day and insufficient quantity.

The fresh surface water resources of the South Caucasus mainly consist of runoff from the Kura–Araz River basins. Kura with its vast river system is the key water provider or, as hydrologists say, is the main water artery of the Caucasus. The river flows through the territories of Turkey, Georgia, and Azerbaijan Republics. The total length of the river is 1,364 km and it has a total watershed area of 188,000 km^2. Of this, 58,000 km^2 are in Azerbaijan, 34,700 km^2 in Georgia, 29,800 km^2 in Armenia, and 66,000 km^2 in Iran and Turkey (Imanov and Verdiyev, 1998; Verdiyev, 1999, 2002; Verdiyev et al. 1999). Twenty-five percent (7.5 km^3) of total water resources of rivers is formed within the Azerbaijan Republic. At present, water resources of Kura and Ganikh in Georgia and of Araz in Turkey, Iran, and Armenia have decreased by 20% as a result of water intake. Also by taking into account water loss from the channel (riverbed) of the Kura River lower course, one may find that water resources of the river available in the Azerbaijan Republic at present is 30% less than should naturally be.

There are more than 10,000 rivers in this territory including small and shallow rivers. Drainage density makes up 0.36 km/km^2 over the Azerbaijan Republic; that is, two times less in comparison with neighboring countries of the Caucasus, where upper parts of basins of transboundary rivers are situated. The river grid in arid areas of Kura–Araz lowland is less developed. Many rivers of Shirvan zone where cities such as Shamakhi, Akhsy, and Sumgait Kurdamir are located, are drying up due to the lack of atmospheric precipitation and a high level of evaporation in summer and do not reach the sea or main river. Water demands of the population in these areas need to be solved

Being a water-poor region, water supply over the Azerbaijan Republic territory totals about 100,000 m^3/km^2, which amounts to an average of about 1,000 m^3 of water per person per year. Accordingly, Azerbaijan Republic occupies one of the lowest places in the world in water availability. Water resources of the Republic are distributed very irregularly over administrative districts. Water resources of Sheki-Zakatala zone, Khachmaz and Kelbajar economic districts exceed those in other areas.

Absheron and Kura–Araz lowlands where such industrial cities us Baku, Sumgait, Alibayramli, and Salyan are located, and where most of the country's population live are the most water-poor regions. During the period of vegetation in the Republic, river runoff amounts to only 10–20% of the annual amount and in Lenkaran-Astara zone it does not exceed 5%. The water supply of the Azerbaijan Republic territory situated downstream of transboundary rivers makes up about 100,000 m^3/km^2. This quotient amounts to an average of approximately 1,000 m^3 of water per person per year.

According to Budiko classification, the territory of country belongs to the insufficient moist climate zone, and especially the lower part of Kura River, which is regarded as an arid zone.

Therefore, it is necessary to redistribute runoff from one part of the territory to another by means of various water economy facilities.

Presently, in the territory there are many water reservoirs providing the population, industry, agriculture, energy, and other sectors with water and to regulating annual and long-term runoff. Reservoirs also help to mitigate negative impact of floods on the quality of water and on damage to water supply facilities. In spite of this limited water resources and the lack of an economical water use scheme, unsatisfied needs for water provision still remain.

2. Major Water Problems and State Water Management Policies in Emergency Situations

The drinking water supply problem has special importance for Azerbaijan, in which water resources are strictly limited. In consideration of the fact that three fourth of its population uses surface waters for drinking and household purposes, and these waters come mainly from the Kura–Aras basin, any pollution and degradation of water in these rivers threaten human health and the environmental condition of the country.

Water provision to the different economy sectors and the population is carried out by means of active hydrotechnical constructions, which are old-fashioned and many water intake facilities and treatment systems cannot operate during high flooding, water turbidity, and extreme pollution. For example, on September 8, 1999, turbidity of the Kura River near the mouth where the water intake facility for provision of Baku city is located, had increased tens of times as compared with previous days and measured 9,817 g/m^3. As a result, the water treatment system could not operate in its normal regime and interruptions occurred in the water supply of Baku and Sumgait cities and the quality of water worsened.

There are also many other cases in which water supply problems have occurred as the result of entrance into the treatment system of highly polluted waters from urban areas, industry, and agricultural lands.

The main directions for effective water provision to the population of Azerbaijan in emergency situations in context of sustainable development, point to the necessity of taking measures for implementation of national and regional programs on water resources management.

Effective steps for improvement of water management in emergency situations include the important need to create and realize a strong legal and institutional basis that provides for the required quality and quantity of drinking water to initiate development projects on determination and use of alternative water sources in the case of an emergency situation and use of modern technology and methods of water provision.

After the South Caucasus countries gained independence, a comprehensive environmental law was passed and new legislation reflecting the important economic and political changes since that time has been prepared. The accepted laws of these countries on Environmental Protection correspond to realized reforms and to international conventions.

Recently, thanks to the financial support of the World Bank and of other international financial structures, the water supply systems in Baku, Ganja, Sheki, and other cities is being renovated and its performance is being improved with the result of a decrease in the number of emergency situations in the water provision of cities.

Generally speaking, the water sector is one of the most important parts of country's economy and engineering municipal infrastructure of the Republic. Central, regional, and territorial governances, joint-stock companies, and private and municipal structures are involved in the water activities. Their performance is regulated by sartorial laws and number of standard acts and bylaws.

Water policy is mainly implemented through water legislation, particularly, through the Water Code of the country – the basic legislative act regulating the water sector. Based on this Code, several Laws, and a number of bylaws were adopted. Particularly, the Law on Water Supply and Wastewaters, Law on Melioration and Irrigation, Law on Hydrometeorological Activities, Law on Entrails, Law on Municipal Water Economies, Law on Safety of Hydrotechnical Installations, etc.

The Water Policy of the country is also implemented through national plans, state programs, and action plans. By the end of 1996, the Government of Azerbaijan, with support of the World Bank, developed a National Environmental Action Plan, which highlights the importance of water resources management and protection. Since the priorities change with time, this National Environmental Action Plan will be periodically revised. Eventually, selection of priority environmental activities shall be carried out with consideration of needs of social and economic development of the country. Presently, the environmental goals of Azerbaijan are focusing on efficient use of natural resources, improvement of health of the population, and prevention of loss of biological resources (UNDP/SIDA, 2005a).

Transition to market economy provides a unique opportunity to consider environmental problems while carrying out economic and political reforms. On February 18, 2003, the Government of the Republic of Azerbaijan adopted a National Program for Sustainable Socio-Economic Development of the Country in the environmental context. One section of the Program is fully dedicated to the Water Policy issues. According to the national program, by 2010, every person in the country shall have access to good quality water. The program aims at introduction of the globally adopted concept of sustainable development to the country.

The program implies development of a special program for rational use of water resources, encourages application of incentives for rational use, improvement of drinking water quality, revision of laws regulating activities concerned with ecosystems, protection of transboundary rivers from pollution, and involving riparian countries in rational use of water resources. In view of this, active dialogues at regional and international levels are being conducted.

Unfortunately, due to financial difficulties of the country in the transitional period, many of the above activities cannot be performed. Therefore, the support of donor countries and international financial institutions is required. In order to support implementation of a national program, the action plan reflecting problems existing in the water sector has been developed and approved.

The program sets out priority directions for protection and rational use of water resources in context of sustainable development, following necessary measures for implementation of national and regional programs on water resources and sites management in emergency situations is required:

- Development and implementation of state programs for efficient use of water resources.
- Stimulation of efficient use of water resources.
- Improvement of water quality and bringing it to the European standard.
- Improvement of legislation in sphere of water ecosystems.
- Application of integrated approaches in transboundary river basin management.
- Improvement of existing water supply and wastewater sites by use of modern technology.
- Use of alternative sources of water and water treatment facilities during emergency situations created by water pollution, floods, or breakdown in water supply systems.

The National Program on Restoration and Enlargement of Forestlands in the Republic of Azerbaijan (2003) also includes water policy principles. The forestation may decrease soil erosion and the number of floods and mudflows and thus, emergency situations in water provision.

In May 2004, the President of the country approved the State Program on Effective Use of Summer and Winter Pastures, Hayfields, and Prevention of Desertification in the Republic of Azerbaijan (2004–2010), which reflects requirements of the Water Policy.

Besides the above-mentioned directives and documents, there are several programs on socio-economic development, which are currently being implemented in the country: State Program on Poverty Reduction and Economical Development of the Country for 2003–2005 (February 20, 2003); State Program on Socio-Economic Development of Regions of the Republic of Azerbaijan for 2004–2008 (February 11, 2004); State Program on Tourism Development in Azerbaijan for 2002–2005 (August 27, 2002). All these programs pay special attention to water issues.

The goal of the State Water Policy is ensuring a safe environment for human beings and satisfying the population's demands. At the same time, the right of other countries or future generations to have safe environment shall not be violated.

The State Water Policy of the country mainly provides for:

- Access of all parts of the population to safe drinking water complying with all quality requirements.
- Balance between economic needs and replenishment ability of water resources.
- Exercise the right of present and future generations to use environmentally valuable water resources.

Mentioned above are the main objectives of Sustainable Water Use, which is the strategic goal of the Water Policy.

Analyses of positive international experience and the current situation allow defining basic principles of the State Water Policy:

- Basin planning and territorial administration of water-related activities.
- Permanent and systematic reduction of negative impacts in water sites.
- Stepwise transition to self-financing of the water sector of the economy.
- Transparency and wide involvement of the society in the decision-making process.

A clearly formulated State Water Policy creates a platform for achievement of agreements on reasonable and equitable use, restoration and protection of transboundary water sites.

It is well known that the population and the economy of the country, and in the first place the agricultural sector, experience lack of quality water. In some regions, particularly on Apsheron Peninsula, there is an acute deficit of water resources.

Specificity of local resources shows that the country is fully dependent on the Kura basin; this is connected to close geographic links with the neighboring countries and imposes certain mutual international obligations.

In practice, water quality degradation can be observed throughout the region, especially in the surface water sites. Only part of the industrial and household waters undergoes treatment procedures: not every town, let alone rural settlements, has treatment facilities, and even those existing are often out of order, and do not provide for the required quality of treatment.

Since most of the water supplied to the population comes, as mentioned, from the Kura basin, pollution of the main rivers of Azerbaijan – Kura and Aras – represents a major problem. Kura and its tributaries are heavily polluted both in the territories of neighboring countries (Georgia and Armenia) and in Azerbaijan as well.

Treatment facilities existing in the country do not comply with modern requirements. It should be mentioned that out of 75 towns and regional centers of the republic, only 35 are equipped with treatment facilities, which are totally deteriorated and perform only mechanical treatment of wastewater.

Due to unsatisfactory performance of treatment facilities in most of the settlements of all three countries of the basin, polluted wastewaters discharge into the rivers and their tributaries. All these pollutants destroy ecosystems of Kura and Aras, causing great damage to the water, environment, and biological systems of the Caspian Sea.

Concentrations of heavy metals (copper, nickel, chrome, etc.) in natural waters exceeds allowable concentrations. Neighboring countries have a significant contribution to the pollution levels. All the above shows in conclusion, considerable pressure on fresh water resources, which affects surface waters (rivers, lakes), causing their eutrophication, degradation, and forming hazardous bottom sediments, and also has an influence on groundwaters – in places of concentrated technogenic load (cities, live stock farms), pollution of deep groundwaters representing the main source of drinking water supply for cities has been registered.

There is no arranged state water management system based on basin principle in Azerbaijan. It is necessary to identify one state authority, which would bear responsibility for providing the population and the economy with water resources, and taking measures for improvement of conditions in rivers and reservoirs. Established practices of setting limits for water consumption and discharge of wastewaters does not consider actual volumes and types of production, technologies, and water supply schemes (direct, reverse, etc.) existing at the enterprises.

Tap water satisfies needs of only 50% of the population, and some areas experience lack of drinking water. Due to the lack of water supply networks and deteriorated conditions of those existing, about half of the water is lost within the distribution system.

In order to resolve problems of rational use of water resources, several important actions are being carried out. In consideration of the acute water deficit per capita in Azerbaijan, in the first place, it is necessary to arrange accurate registration of consumed water. Introduction of a program on reduction of water losses in city zones is the top priority issue.

One of the most urgent problems of the country, from the point of view of rational use of water resources, is sanitary protection, which will be directed towards implementing control of natural and drinking water quality, harmonizing all standards with international norms and creating a system of national standards. Water protection and water resources management measures shall be included into strategies of all sectors of the economy in the country.

In the medium- and long-term perspective, water resources shall be protected through investing in construction of cities' sewage systems and industrial treatment systems along the Kura River.

The water sector – one of the most important parts of the nature, environment, social, and economic life – requires permanent attention of the state. According to the National Program on Socio-Economic Development, priorities of the Republic of Azerbaijan in the water sector are protection of the environment and achievement of rational use of natural resources. This program, being an integral part of the development strategy of the country, is aimed at coordinating national and regional efforts for protection of the environment, implementing scientifically based development principles, ensuring sustainability in use of economic and human resources for present and future generations. One of the most important trends also is coordinating efforts of government and nongovernmental organization (NGO) sectors for protection of the environment. Despite the fact that water problems are reflected in the mentioned State Program and National Concepts of Social and Economical Development, and Protection of Environment, in our point of view, Azerbaijan should develop a separate national concept, which will cover specifically water issues (in other words – National Water Policy).

In the conditions of market economy, in order to accelerate resolution of the water sector problems in the country, ensure its operative management, law enforcement, observance of international conventions on water-related issues, international and interstate cooperation, are necessary to create a State Commission on Water Issues.

3. Organizations Responsible for Water Supply of Cities and Entire Country: Their Strengths and Weaknesses

There is a complete water utilization system in the country, constructed during the Soviet times. Water provision to the economy sectors and the households is carried out by means of active hydrotechnical constructions, which enable the redistribution of surface waters from one part of the territory to another and according to the seasons of the year.

Organization of the management of the water resources of Azerbaijan is carried out in accordance with existing legislation. The laws adopted by Parliament are enforced by relevant decrees of the President of the country. Among these decrees there are such that determine the authority of relevant state bodies. The Ministry Cabinet of the country in accordance with the decrees of the President adopts a number of necessary regulations and decisions prepared by corresponding bodies of executive power (UNDP/SIDA, 2005b).

The following organizations are dealing with the issues of water resources management in Azerbaijan:

- Ministry of Ecology and Natural Resources.
- Committee of Amelioration and Water Economy[2].

While developing plans and programs in the field of protection and use of water resources, the two above-mentioned bodies engage the following other agencies:

- Ministry of Fuel and Energy in the issues of water use for energy purposes.
- Ministry of Agriculture.
- "Azersu" Joint-Stock Company.
- Ministry of Health.
- Water User Associations.

The following organizations take part in research and other activities in the field of protection of water resources:

[2] After this article has been delivered some structural changes took place. Under the decree # 467 of the President of Azerbaijan Republic, dated October 23, 2004, the Committee of Amelioration and Water Economy was included as an Agency into structure of the Ministry of Agriculture.

- Institute of Geography of the Academy of Science.
- Baku State University.
- National Committee on International Hydrological Program of UNESCO.
- Environmental NGOs.

Furthermore, the organizations responsible for management and protection of water resources, in cooperation with NGOs in this field, inform and involve the public in the decision-making process.

Activities of the State Water Protection Organization encompass the following fields of water resources management:

- Drinking and industrial water supply, irrigation.
- Reservoir management and regulation of runoff.
- Discharge of wastewater.
- Control and monitoring of water quality.

Regulation of development, coordination, state examination, approval and implementation of schemes of integrated use and protection of water resources is established by the Ministry Cabinet of the Azerbaijan Republic. State programs for use and protection of water bodies are developed by the Ministry Cabinet of Azerbaijan Republic with consideration of proposals of the Committee of Amelioration and Water Economy, Ministry of Ecology and Natural Resources, municipalities, users of water bodies, and social organizations, in a procedure established by law.

State monitoring of water bodies is carried out by the Ministry of Ecology and Natural Resources and the Committee of Amelioration and Water Economy in the procedure established by legislation of the Azerbaijan Republic. In Azerbaijan, the state accounting of waters is conducted by uniform system in the procedure established by the Ministry Cabinet of the Azerbaijan Republic and is based on accounting information from users of water bodies and state monitoring. The State Water Cadastre is kept by the Ministry of Ecology and Natural Resources and Committee of Amelioration and Water Economy in a procedure established by legislation of the Azerbaijan Republic.

State ecological examination carried out with the purpose of definition of compliance of construction projects and reconstruction of industrial and other facilities having an impact on the condition of water bodies and their compliance with established standards, technical conditions, and requirements is conducted by the Ministry of Ecology and Natural Resources and Ministry of Health of the Azerbaijan Republic.

Regulation of state control and standardization of use and protection of water bodies are approved by the Ministry Cabinet of The Azerbaijan Republic. Regulations of standardization in the field of use and protection of water bodies are established by the Ministry Cabinet of The Azerbaijan Republic. The right for use of water bodies can be obtained by the procedures dictated by legislation of The Azerbaijan Republic. The President establishes the types of use of water bodies, which require special permission. The state control on the observation of regimen of use and protection of natural resources and other economic activity in water protection zones is conducted by the Ministry of Ecology and Natural Resources and Committee of Amelioration and Water Economy within the limits of their authority.

Control of water protection zones is conducted by the Committee of Amelioration and Water Economy at the Ministry Cabinet of The Azerbaijan Republic, which have the right to stop or prohibit operations having harmful impact on the condition of water bodies. Control of protection of groundwaters is conducted by the Ministry of Ecology and Natural Resources. Admissible norms of harmful influence on water bodies are established by legislation on environmental protection and water and are approved by the Ministry Cabinet of The Azerbaijan Republic.

Volumes of minimal flow of water and water intake without regeneration are established for each water body by the Committee of Amelioration and Water Economy at the Ministry Cabinet of The Azerbaijan Republic and Ministry of Ecology and Natural Resources. Basin agreements are completed between the Committee of Amelioration and Water Economy at the Ministry Cabinet of The Azerbaijan Republic, municipalities, and water users. Basin agreements are prepared on the basis of water balances, schemes of integrated use and protection of water resources, state programs and other projects of use, restoration and protection of water bodies by proposal of the Committee of Amelioration and Water Economy and Ministry of Ecology and Natural Resources.

According to the regulations, the rules of user charges are established by the Ministry Cabinet of The Azerbaijan Republic. When calculating the charges, the price of water service is taken as the basis. Payment for water consumption from water bodies in municipal ownership is established by municipalities in a procedure mentioned in the first part of this article. The fee for water consumption from water bodies in private ownership is established on the basis of agreement between the owners and the consumers. Types of payment related to consumption from water bodies and rules of payment are established by the Ministry Cabinet of The Azerbaijan Republic.

Authorities of "relevant agencies of executive power," anticipated by Part Four of Article 227 of the Code of The Azerbaijan Republic on Administrative Violations, are fulfilled, respectively, by the following agencies:

- Misuse of drinking water – the Committee of Amelioration and Water Economy at the Ministry Cabinet of The Azerbaijan Republic.
- Issues related to municipal water bodies – Joint-Stock Company "The Azersu".
- Violations of requirements for sources of drinking water, water treatment facilities, sanitary-protection zones of water pipes, noncompliance of quality of drinking water with the sanitary requirements and standards – the Ministry of Health of The Azerbaijan Republic.
- Irrigation of green plantations and washing of transport facilities with drinking water, leakage of drinking water or of domestic wastewaters – local executive municipal bodies.

For the regulation of the provision of the population, enterprises, institutions, and organizations with high-quality water in necessary quantities, meeting the requirements of state standards, discharge of wastewater in accordance with the Law on Water Supply and Sewage, the local executive municipal bodies acting as water supply and sewage system institutions shall follow the following principles:

- Cost recovery of water supply, sewage, and waste disposal services.
- Provide consumers with water of adequate quality and in required quantities.
- Rational use of water resources.
- Creation of reliable systems of treatment and discharge of sewage and waste disposal.

Corresponding executive bodies in agreement with municipalities, in accordance with conditions foreseen by legislation, issue permits for the following:

- Use of surface and groundwaters in defined quantities.
- Discharge of sewage in different surface and groundwater bodies, also transportation and waste disposal (including liquid waste) in soil or water bodies.

- Construction of dams, water reservoirs, and other hydrotechnical facilities.

The plans for the location and development of water supply system and sewage are prepared and approved for the whole territory of The Azerbaijan Republic in a procedure, established by the Ministry Cabinet of The Azerbaijan Republic. Water supply enterprises determine the volume of water used by consumers by means of water meters or by means of calculations in a procedure defined in relevant normative-legislative acts.

The main duties of organizations of the sewage system include:

- Transportation, treatment, neutralization of wastewater, and its discharge into the environment and water basins.
- Expansion and reconstruction of sewage facilities of general use, their technical maintenance.
- Adoption of measures on removal and preliminary prevention of leakage of wastewater from sewage systems into the environment.

Directing of wastewater into water bodies is carried out according with a procedure determined by law. In case of facilities lacking sewage treatment systems the local authorities (municipalities) or the owners of the facilities, should address the organization in charge of sewage and require installation of sewage pipelines or the construction of sewage treatment facilities at their expense.

Water supply and sewage organizations can offer in the agreement the following types, terms, and sums of payment for services for all categories of consumers:

- Payments for the quantities of water received in accordance with active regulations on the basis of information from water meter appliances or calculations, and discharged wastewater.
- Payments for connection to the systems of water supply and sewage.
- Single payment for connection to the systems of water supply and sewage.
- Advance payment for the installation of technical means, including water gauge appliances.

As mentioned above, state management of the use and protection of state-owned water bodies is the responsibility of the Committee of Amelioration and Water Economy at the Ministry Cabinet of The Azerbaijan Republic and Ministry of Ecology and Natural Resources within the limits of their authority. For developing plans and programs in the field of protection and use of water resources, other agencies are engaged. For example, the schemes of integrated use and protection of water resources are developed by the Ministry of Ecology and Natural Resources, Committee of Amelioration and Water Economy, JSC "Azersu," and other organizations. Following is information regarding several of these organizations and their activities.

3.1. THE MINISTRY OF HEALTH

The Centre of Epidemiology and Hygiene is responsible for setting the standards and monitoring drinking water quality.

The Laboratory of the Republic Sanitary Service of the Ministry of Health is certified by the State Committee of Standards. During the past few years, the laboratory has been cooperating with the International Red Cross and Physicians Without Borders in monitoring water quality. Results of the analyses were controlled in Great Britain. Through financial support of the World Health Organization, during 1999, 42,688 samples of water were analyzed. The laboratory participated in the preparation of a report on water quality in Azerbaijan.

Problems: absence of modern equipment for measurement of radioactive emanation and absence of reagents. Unfortunately, water standards in the country are very strict and do not comply with standards of European Union (EU).

3.2. "AZERCU" JOINT-STOCK COMPANY SECOND-VALUE HEADING

The "Azersu" Joint-Stock Company was established in July 2004, taking over the issues of water supply of the cities of Baku and Sumgait which were formerly managed by Absheron Joint-Stock Company as well as the water supply and sewage services of other regions of the country which, until then, were managed by the State Committee on Architecture and Civil Engineering. The basic function of "Azersu" is operation and rehabilitation of water supply and sewage systems. The Company includes also a research and planning institute "Water-Canal."

Previously "Azersu" employed round 2,500 workers. After the reform their number almost doubled. The subdivision of the Company dealing with financial issues coordinates the activities of selling water and installation of water meters. It has central and regional laboratories for analysis of the quality of drinking water.

"Azersu" JSC established different tariffs on the use of water by residents, organizations funded from the budget, and by industry (185 manats/m^3 or \$0.04/m^3, 800 manats/m^3 and 2,200 manats/m^3, respectively). Owing to economic problems, lower tariffs were set for population (realistic tariffs are close to 500 manats/m^3). There are few water meters. The collection of bills is 80%. The collected funds do not cover the costs.

By financial support of the World Bank and the European Bank of Reconstruction and Development (\$86 million) and the Government (\$10 million) the Project on Reconstruction and Rehabilitation of Water Treatment Facilities of Djeiranbatan and Kura Water Pipelines has been implemented (partially). Through a grant from the Government of Sweden, 370 of the 900 pumping stations of the city of Baku have been repaired and grouped. Through financial assistance of the World Bank (\$13.5 million) the project on partial reconstruction and rehabilitation of water treatment facilities, water intakes and others, damaged in the earthquake in 2000 has been implemented.

Problems:

- Because of low salaries, work motivation is low and highly skilled professionals go to other organizations for higher paid jobs.
- There is a lack of technology experts.
- There is the need for training of human resources in accordance with international standards, including training in corresponding institutions of education of developed countries.
- Available funds are insufficient for ensuring normal functioning of the system.
- Large funds for maintenance and reconstruction of water supply and sewage systems are required.
- Water meters and modern laboratory equipment is required.
- It is necessary to provide the "Water-Canal" Research Institute with material-technical and methodological support.
- To raise the effectiveness of their management of water supply and sewage systems, modern computer technology and computer-based systems are required.
- Study of international experience about water supply and sewage in correspondence with EU directives is required.
- Because of absence of water meters water use is inefficient.
- The water treatment facility in Govsany is in very bad condition (for this the Government of France intends to allocate credit of \$35 million).

- Funds are needed for the reconstruction of almost the entire system of water supply and sewage and construction of new treatment facilities (at least one in Lokbatan for the city of Baku) are required.
- Quality of water supplied to the population from the Kura does not meet the requirements of the State Standard 2874–82 "Drinking Water." Because of the absence of reagents the system often cannot function at full capacity. This is due to the above-mentioned fact that surface sources of water are heavily polluted while the treatment facilities are deteriorated (for renovation and reconstruction of which financial resources are required). Because of nonsatisfactory conditions of the distribution network there is leakage of water and polluted agricultural, domestic, and industrial waters flow into the system.
- The sewage system of the city of Baku covers only 70% of its territory and only about half of sewage is treated (40% biologically, 40% mechanically).
- Owing to rapid growth of turbidity of Kura (and its inflows) during high water the water treatment facilities are rendered inoperable thus causing failures in the water supply of the population of the city of Baku. Such situations mainly take place in autumn and spring on the average 3–5 times a year for 1–2 days. In the system of centralized water supply of the city of Baku about 300 emergency cases occur annually, caused mainly by nonsatisfactory conditions of the water distribution network, period of validity of which has expired long time ago.
- Practically nobody works with the population to promote efficient water use practices.

3.3. THE MINISTRY OF FUEL AND ENERGY

The Ministry of Fuel and Energy is responsible for issues of water use for energy purposes. During the operation of large water reservoirs, coordinated actions with the Committee of Amelioration and Water Economy, which is concerned with water intake in Upper-Karabakh and Upper-Shirvan canals, are necessary.

3.4. LOCAL EXECUTIVE BODIES

These bodies are responsible for carrying out the policy of treatment and supply of the population with good quality drinking water and for coordinating the development of standards for water use.

3.5. SCIENTIFIC RESEARCH ORGANIZATIONS AND RELEVANT INSTITUTIONS OF HIGHER EDUCATION

Baku State University, the University of Architecture and Construction of Azerbaijan, and the Academy of Oil of Azerbaijan train specialists in the field of water resources, at the levels of masters and postgraduate students. Scientific research is being carried out. Due to the small scholarships at the universities and salaries at governmental organizations, lack of modern educational technologies and laboratories, the knowledge level of graduating students is not very high.

3.6. NONGOVERNMENTAL ORGANIZATIONS

Major NGOs are the International Hydrological Program under UNESCO, ECOREC, RUZIGAR, and SANIA, which carries out the projects on providing information to the public regarding the problems in the water sector; explaining to the legislators the legal aspects of water resources protection issues through publishing bulletins, booklets, press articles and training.

Due to the lack of financing for the implementation of works in the water sector, NGOs carry out activities in the framework of projects financed through grants. More often these projects are directed at the raising the awareness of the population. The existing system of dissemination of environmental information, unfortunately, does not fully allow the NGOs to obtain desirable environmental information. Independent sources are very limited. Sometimes NGO representatives have to conduct the monitoring of water quality themselves.

3.7. LABORATORIES

There are the following independent laboratories carrying out water quality analysis in the country:

3.7.1. Laboratory – Ecolab

Laboratory – Ecolab carries out work connected with sample taking, chemical, and microbiological monitoring of surface and groundwaters, and bottom deposits, etc. It has participated in various projects financially supported by the World Bank, European Bank for Reconstruction and Development, and other international organizations.

3.7.2. Research Centre Laboratory "Spectrum"

The laboratory has up-to-date equipment and was completed with the help of TACIS. The laboratory employs highly qualified specialists and has the certificate of the State Committee of Standardization of the Republic of Azerbaijan. It cooperates with different international organizations.

3.7.3. ERT Environmental Caspian Laboratory

The objective of the laboratory is to carry out the monitoring of marine water and soil samples, etc. Equipped laboratories operate in Azerbaijan and Scotland.

3.8. DONORS

Different projects on water were and are being implemented in the region financed by the World Bank, EU TACIS, USAID, UNDP/GEF, EBRD, IDB, SADC, and GTZ/KfW. Often these organizations coordinate their activities between themselves. Within the framework of USAID–DAI and Tacis JRM meetings were arranged with the participation of the representatives of water protection establishments. USAID–DAI created web pages for the Hydro-Mets of the countries in the region and developed the database on the quality and quantities of river water for each of them (USAID, 2002). This site allows the population to have access to information on water resources. TACIS JRM carried out water quality monitoring, standards, selection of hot spots, rehabilitation of laboratories, and others.

The World Bank project is directed at the creation of the basis for the regulation water supply, irrigation and drainage systems, etc. Kreditanstalt für Wiederaufbau (KfW) and Asian and Islamic Banks of Development are financing projects on the reconstruction of water supply and wastewater systems, fighting mudflow effects and others. Coordination of interactions between the projects and governments is the key factor for the achievement of water resources efficient management.

On February 28, 2003, in Baku a workshop was arranged by OSCE and USAID on the issues of national priorities in the water sector. Among the participants were 13 representatives of different organizations engaged in the activities related to the water resources policy and management in Azerbaijan. The meeting focused on the discussion of water-related problems from national viewpoints with the purpose of revealing priority issues for the solution of problems in water sector. These issues were also considered from the regional viewpoint taking into account their particular importance for Azerbaijan.

The workshop participants discussed the following issues:

- General situation in the water sector.
- Juridical, economic, social, and environmental issues of water use and water protecttion.

A Regional Workshop on Priority Issues in the Water Sector with the participation of specialists of the region held in May 2003 in Tbilisi was arranged by the same organizations and emphasized the importance of a number of reforms in the water sector in the region.

4. Identification of Needs for Improved Water Management in Azerbaijan

Having taken into account the results of the analysis on the present state of water supply and the above-mentioned problems and proposals, as well as the opinions of specialists expressed during the meetings arranged by OSCE, USAID, UNDP, and the results of this author's report prepared for projects of UNDP SIDA (UNDP/SIDA, 2005a, b) and USAID (USAID, 2002, 2003) the following recommendations are proposed for the improvement of the water sector institutional and administrative aspects both at national and regional levels.

4.1. NEEDS AT THE NATIONAL LEVEL

- Improvement of institutional structures of the organizations engaged in water sector in compliance with the structures established in EU countries.
- Construction of modern treatment facilities and creation of their operation system.
- Reconstruction of water supply and sewage systems.
- Reorganization of the existing system of standardization and establishment of maximum allowable discharge of wastewater flowing into the rivers in compliance with the EU Directives.
- Reorganization of the existing system of standards for water use and wastewater discharge in compliance with the EU Directives.
- Reconstruction of sewage system of the city of Baku and other big cities.
- Development and adoption of integrated water quality and quantity management gears on the basis of basin-wide approach. With regard to the transboundary rivers, develop and introduce the integrated river basin management plans in cooperation with other countries of the basin.
- Establishment of State Water Commission and development of a national plan and strategy on water resources, creation of relevant structures for their implementation.
- With the purpose of preventing parallel activities in the surface water quality monitoring, make necessary changes in the controlling bodies of the organizations.
- Introduction of computer-based systems into monitoring and data exchange systems.
- Introduction of computer-based systems into water supply and sewage systems.
- Increase in salaries of the specialists and application of encouragement methods for the attraction of people to work in water sector.
- Amendment of the fining system for the cases when the discharged wastewater does not meet established standards, in accordance with the EU Directives.
- Improvement of the tariff systems for the use of the water by the population, state budgetary organizations, and industry.
- Preparation of priority requirements for investments.
- While assessing the capabilities of the laboratories requirements for trainings should be revealed and the program for meeting these requirements should be developed. Training of the specialists in water resources management should be conducted on the basis of the principles stated in the EU Directive in the field of water policy and in Helsinki Convention.

- Conditions for providing all interested parties with the information obtained and prepared in the framework of the project implementation should be created. Make and maintain web page, which would be used for obtaining and distribution of information on the project. Preparation of simple advertising booklets in the national language containing the information on the project and disseminated at the request, should also be included.
- Water protection organizations and water users through the installation of water meters, conducting the work with the population about economical water use, and through the irrigation system reconstruction should fight against the existing high losses of water in water supply systems with combined efforts (together).
- Ministry of Ecology and Natural Resources and other relevant organizations should try to achieve the real reduction of the discharges of untreated wastewater into the water bodies and rehabilitation of water treatment plants. Industrial enterprises should ensure the relevant treatment of wastewater before their discharge into the water bodies.
- For the purpose of more efficient work of national water legislation, it is necessary to make relevant amendments to the legal acts (including the status, instructions, taxation, and water use licensing).
- Increasing the role of water users in water resources management, expansion of water users associations and ensuring the adherence to international standards.
- Booklets on the water resources, on the conditions existing in water use and water resources protection should be published.
- Organizations engaged in the water sector shall work out and submit proposals on the improvement of water supply system and wastewater treatment, on the construction of installations (facilities) for circular water supply, and project proposals related to the use of modern methods and technologies in the field of irrigation.
- Development of payment systems.
- Providing relevant information to the population on sanitary and epidemiological state of waters.
- Improvement of public participation in water resources management.
- Rehabilitation and development of monitoring network, creation of database in relevant format and ensuring the data exchange between the organizations.
- Preparation and publication of Water State Cadastre.
- Development of informing system in emergency situations.
- Taking into consideration the economic and political situation created in the region, render support to international organizations and donors for the rising of the level of cooperation in the region.
- For the realization of new approaches in water resources management, foreign assistance including the assistance of international financial institutions is required for Azerbaijan. At the same time, at the initial stage some pilot projects in separate river basins and big water economy units, as well as in the sphere of the improvement of legislative acts should be implemented.
- At the national level, taking into account the market economy and new conditions of land husbandry gradual transition to the integrated management should be carried out under the leadership of the Ministry Cabinet and the Parliament of the Republic. At the same time, the key players in this initiative should be the Committee on Land Reclamation and Water Industry and the Ministry of Ecology and Natural Resources and subordinate to them scientific and research and designing institutes. For the purpose of the development and implementation of the National Plan (IMWR), National Academy of Science and other organizations engaged in the solution of water issues, as well as public should be attracted.

4.2. NEEDS AT THE REGIONAL LEVEL

- Carrying out operational and routine (regime) information exchange on the quality and quantity of surface waters, creation of early (timely) notification system.

- Arrangement of international trainings and workshops on monitoring and data exchange.
- Creation of an international monitoring group of experts, and carrying out the inventory of the sources of water resources pollution through the involvement of local independent experts for the purpose of developing priority measures for waters protection.
- Ensuring close cooperation between the ongoing regional projects in the field of water resources.
- Introduction of common standard operational procedures of water quality and quantity monitoring.
- Meteorological organization.
- Seeking the possibilities for obtaining and restoring the hydrological and hydrochemical statistical data from the funds of Russia and their conversion into the created database.
- Arrangement of joint expeditionary researches in emergency situations.
- Social and economic conditions do not allow the population to pay for water use, but the problem could be temporarily mitigated through state subsidies.
- Improvement of the water-related legislation on the basis of the principle of basin-wide management.
- Carrying out stage by stage institutional and organizational reforms.
- Establishment of river basin management council.
- Preparation of National Water Program, development of National Water Policy and Strategy.
- Coordination of projects and programs that are being implemented in water sector.
- Application of modern technologies for wastewater treatment, which are rather simple to be used, and, at the same time, requiring not much energy for operation.
- Improvement of the monitoring system of the Ministry of Health at the sources of drinking water supply.
- Harmonization of the norms and standards in water sector.
- Arrangement of courses for the professional skills improvement of a wide range of specialists in water sector.
- Ratification of international agreements on water issues.
- Enforcement of European standards of water quality and quantity monitoring.
- Harmonization of water cadastres preparation (drawing up, working out) on the basis of common European system.

References

1. Verdiyev, R.H. (2002) Water Resources of the Eastern Caucasus Rivers Under Climate Change, Monograph, Baku, 230.
2. Imanov, F. and Verdiyev, R. (1998) Estimation of changes of natural and ecological run-off of Kur River for different climate change models. Ecological problems of Caspian Sea and ecological education in Caspian countries, November 18–20, 1998, Baku-Azerbaijan, p. 62.
3. Verdiyev, R. (1999) Water resources assessment for the river Kura in conditions of climate change, World Meteorological Organization Bulletin, Genevre, pp. 327–328.
4. Verdiyev, R., Mansimov, M., Imanov, F. (1999) The role of meteorology in economy of the Azerbaijan Republic, Thesis of presentation, IVth European Conference on Application of Meteorology, Sweden, September 13–17, 1999.
5. The World Bank (1993) Executive Summary, Water Resources Management, which describes the Bank's framework for policies and institutional reforms, http://lnweb18.worldbank.org/essd/essd.nsf/WaterResourcesManagement/ExecutiveSummary
6. For a useful institutional guide, see annotated list of government agencies responsible for water resources management, which covers all three countries, Annex 4, Strengthening water management in the South Caucasus, Final report, Development Alternatives, Bethesda, Maryland, April 2001.
7. National Environmental Action Plan (1998) State Committee on Ecology and Control of Natural Resources, Baku, p. 53.

8. DAI (2002) International Experience in water management. Baku, January 2002.
9. Priority issues in water sector (2003) Materials of Regional meeting OSCE/USAID, Tbilisi, May 16, 2003.
10. UNDP/SIDA (2005a) Reducing transboundary degradation of the Kura–Aras River Basin, Water policy in Armenia, Azerbaijan and Georgia, Task report, Tbilisi.
11. UNDP/SIDA (2005b) Reducing transboundary degradation of the Kura-Aras River Basin, Institutional aspects of water resources management in Armenia, Azerbaijan and Georgia, Task report, Tbilisi.

6.2 UNIFICATION OF STANDARDS FOR DRINKING WATER FROM DIFFERENT SOURCES: PRELIMINARY APPROACH FOR WATER SECURITY IN ISRAEL

TZVI LEVINSON AND GIL DROR
The Levinson Environmental Law Firm, Haifa, Israel

Abstract: According to the existing legal status, drinking water, which is not mineral water or spring water and supplied in bottles is not regulated under the People's Health Regulations (sanitary quality for drinking water). It is not regulated under the People's Health Regulations (mineral water and spring water), either, although both kinds of water are supplied in bottles, which makes them having the same marketing characteristics that may influence the quality of the water (e.g. the type of the bottle, amounts and types of the bacterial population). The Israeli standard setting standards for the quality of drinking water, which is not mineral or spring water supplied in bottles and containers, is not mandatory. Thus, the public does not have any certainty as to what the standards of drinking water in bottles should be. In cases where the main water supply will be restricted or unavailable due to security reasons, the most likely scenario will be the supplying of water in bottles or large containers It is vital therefore that this method of supply be regulated to avoid large-scale diseases. We therefore recommend, to adopt the Israeli standard as mandatory for drinking water supply in bottles or alternatively, to unify the standards of drinking water regardless of the method of supply.

Keywords: water security, water standards, water health regulations, water legal status

1. Introduction

Improving the security of the national resources of drinking water has become a top priority since the events of September 11. Significant actions should be taken to assess and reduce vulnerability of the drinking water supply to potential terrorist attacks. To be prepared for an attack on national reservoirs of drinking water it is important to practice how to respond to emergencies and incidents, to develop new security technologies, to detect and monitor contaminants and to avoid any shortcomings, which can be prejudicial to security.

However, the development of strategies for the security of drinking water on the national level should be based on modern and well-established standards for drinking water – standards that should be the determined as the base line for protection.

Therefore, the determination of uniform standards for the quality of drinking water from different sources is necessary for further discussions on water security. The determination of uniform standards for the quality of drinking water indicates the importance that the state attaches to the supply of high-quality drinking water in comparison with other tasks such as protection of the environment, protection of public health, and education.

Most of the drinking water in Israel is supplied by water suppliers through pipes to households. However, an additional important source of drinking water is the water pumped from wells or springs and supplied in bottles or containers to the consumers and therefore, it is important to set adequate standards applicable to drinking water supplied in bottles and containers. In case of a terror attack, this source of water may provide a useful solution in the short run.

In this article, we will review the Israeli legislation concerning quality and standards for drinking water. We will present comparative data on the standards for drinking water supplied

Y. Sharan et al. (eds.), Water Supply in Emergency Situations, 159–163.

through pipes by the public drinking water supply and drinking water supplied in bottles, and we will discuss the findings.

2. Definitions of Drinking Water in the Israeli Law

According to the *Public Health Ordinance* 1940, drinking water is water for drinking or for cooking, and in the food industry – water that will or may be in contact with food or with other materials that will be in contact with food. This is a broad definition that does not, however, set any quantitative standards to be met by the suppliers.

A definition of drinking water can also be inferred from the following definition of nonpotable water contained in Article 2 of the *People's Health Regulations (sanitary quality of drinking water)*, 1974 (herein after – the *Sanitary D.W Regulations*): water that is excluded from being drinking water based on microbial tests of *Escherichia coli*, chemical tests of various parameters, high concentration of fluoride, negative findings of sanitary survey, radioactive tests, toxic substances tests, or any other reasonable concern that the water may endanger public health. This piece of legislation is the framework that regulates the supply of drinking water to the public in Israel.

According to the *People's Health Regulations* (*mineral water and spring water*), 1986 (hereinafter – the *Mineral Water and Spring Regulations*) the term "drinking water" is defined as water that is drinking water according to the Sanitary D.W Regulations, with certain exceptions of substances being listed in the annexes to the Regulations.

Last but not least, the Israeli Standard Institute issued a standard for the quality of drinking water in bottles and containers No. 1501, 1994. This relates to drinking water supplied in bottles, which is neither mineral nor spring water. This standard is not legally binding. However, water suppliers that supply water according to the standard can mark the bottles with the standard mark approved by the Israeli Standards Institution.

A comparative analysis of these three standards for drinking water clearly shows inconsistency among them. In some cases, one or more of these standards ignore certain parameters and in other cases different values are set for the same parameter.

3. Results

Table 1 compares the upper limit values of the People's Health Regulations (sanitary quality of drinking water), 1974, the People's Health Regulations (mineral water and spring water), 1986, and Israeli Standard 1501, 1994, only regarding the parameters where inconsistency is found.

4. Discussion

Without referring to any technological, scientific, or engineering aspects regarding the determination of standard for drinking water, from the table above, the following questions arise:

1. What are the binding standards for drinking water, in the light of the different standards in the different pieces of legislation, as described above?
2. Is it justified to set different standards for drinking water supplied through pipes by the public supply system than standards for drinking water supplied in bottles and containers?

As regard to the first question, the standards for drinking water supplied through pipes by the public supply system to the citizens in Israel is regulated by the People's Health Regulations (sanitary quality of drinking water), 1974.

TABLE 1. Summary of different parameters for drinking water in Israel

	Parameters	Upper concentration limits determined by People's Health Regulations (sanitary quality of drinking water), 1974 (mg/l)	Upper concentration limits determined by People's Health Regulations (mineral water and spring water), 1986 (mg/l)	Upper concentration limits determined by Israeli Standard 1501, 1994 (not official) (mg/l)
1	Nickel	0.05	0.05	No value
	Silver	0.01 (0.08 in treated water)	0.01 (0.08 in treated water)	No value
	Taste and odor	Not repellent	No limitation	No value
	TSS	1,500	No limitation	No value
2	Nitrites	No value	0.005	0.005
	Sulfides	No value	0.05	0.05
	Mineral oil	No value	0	Not detected
	Borates	No value	50	50
	Threehalomethanes	No value	0	Not detected
	Pseudomonas aerugenosa bacteria	No value	0[b]	0[b]
	Sulfate-reducing bacteria	No value	0[b]	0[b]
	Streptococcus bacteria	No value	0[c]	0[c]
3	Total β beam activity	1[e]	1.1[e]	1.1[e]
	Cadmium	0.005	0.01	0.01
	Lead	0.01	0.05	0.05
4	Nitrate	70	45	45
5				
	Cyanide	0.05	0.01	0.01
	Color	15[d]	5[d]	5[d]
	Fecal coliform	1[a]	0[b]	0[b]
	Escherichia coli	3[a]	0[b]	0[b]
	Sulfate	437.5 (minus the concentration of magnesium multiplied by 1.25)	200	No value
	Chloride	600	200	No value
	Surfactants	1	0	Not detected
	Copper	1.4	1	1
	Magnesium	150	50	No value
	Phenol	0.002	0	Not detected
	Chlorine	0.5 < chlorine <0.1	0	0
	Organic substances	VOC Benzine - 0.01 Benzo(a)pyrene – 0. 007 1,-2 Dichlorobenzene – 1 1,-4 Dichlorobenzene – 0.3 1,-2 Dichloroethane – 0.05 1,-1 Dichloroethylene—0.03 1,-2 Dichloroethylene—0.1 1,1,-1 Trichloroethylene - 0.2 Trichloroethylene - 0.05 Tetrachloroethylene 0.04 – Chloroform - 0.1 Carbon Tetrachloride - 0.005 Monochlorobenzene – 0.3 Formaldehyde-0.9 Toluene – 0.7 Xylenes - 1 0.05 – סטירן	3	3

	Pesticides Ethylene di bromide – 0.00005 Lindane – 0.002 Alachlor – 0.02 Heptachlor – 0.004 Chlordane – 0.002 Methoxychloride – 0.02 Andrin – 0.002 Atrazin – 0.002 Dibromo(1,2,3) chloropropane – 0.001

1 – Parameters that contain upper limit values in the Sanitary D.W Regulations, but with no reference in the People's Health Regulations (mineral water and spring water), 1986
2 – Parameters that contain upper limit values in People's Health Regulations (mineral water and spring water), 1986, but with no reference in the Sanitary D.W Regulations
3 – Parameters that have stricter values in People's Health Regulations (sanitary quality of drinking water), 1974, than in the People's Health Regulations (mineral water and spring water), 1986
4 – Parameters that have stricter values in People's Health Regulations (mineral water and spring water), 1986, than in the Sanitary D.W Regulations
[a]Number of microbes in 100 ml of water
[b]Number of microbes in 250 ml of water
[c]Number of microbes in 50 ml of water
[d]Platinum cobalt units.
[e]Bq/l
[f]Nephelometric

Regarding water supplied in bottles or containers, we will refer separately to water defined as mineral water or spring water, and to other water.

1. Mineral Water and Spring Water

In the People's Health Regulations (mineral water and spring water), 1986, mineral water and spring water can be used as drinking water under the following cumulative conditions:

1. It was taken from an underground source or spring approved by the competent authority.
2. The connection between the underground source or the spring and the filling of the bottles should be direct, without stops or any intermediate system.
3. The results of the microbial tests do not exceed the values in Annex I.
4. The results of the chemical tests do not exceed the allowed values.
5. The water was not treated.

Treatment is defined in these regulations as: every treatment "except processes of filtration and precipitation." This definition is problematic: on one the hand, mineral water and spring water will be considered as drinking water only if it was not treated, but on the other, the definition of treatment allows filtration and precipitation, which are general terms for most of the common methods for water treatment. Moreover, if filtration and precipitation are allowed, obviously some installation will be located between the source of the water and the filling point of the bottles, which is in contradiction to requirement no. 2 above.

2. Other Water Supplied in Bottles

As regard to water that is supplied in bottles and containers but pumped from wells: since this water is treated before being filled up in bottles, it cannot be defined as mineral water or spring water. This water cannot be regulated under the Sanitary D.W Regulations, since these regulations refer to water supplied through pipes. It seems that the most adequate standard for this water is set in the Israeli standard 1501 that refers to the selling of mineral water as well as of other water in bottles. The problem is that this standard is not official and therefore is not legally binding according to the Israeli Standards Law, 1953.

Article 1 to the Israeli criminal law states that there is no penalty without a law (*Nulla poena sine lege)*, meaning that one cannot be penalized for doing something that is not prohibited by law. This principle ensures certainty of the law and includes the requirement of clarity, e.g. that the law will be understandable. In our case, there is an inherent contradiction in the regulation which may cause the following problems: firstly, lack of legislation that will authorize the administrative authority to oblige the water supplier to meet high standards; and secondly, a standard which is not unequivocally defined can be interpreted in different ways. In this case, regardless how the supplier interprets the law and how he proceeds, he may breach the law and be exposed to legal proceedings against him. Moreover, not only the worker who performs the water supply is exposed to legal proceedings, but also the corporate officers have responsibility to control and to perform all necessary steps to avoid violations of the law. In light of the principle *Nulla poena sine lege,* there is a lack of legislation regarding drinking water supplied in bottles or containers, which is not mineral or spring water.

As to the second question, the upper limit values for drinking water in the regulations and the Israeli standard are inconsistent: in some cases the parameters in the regulations for sanitary drinking water are stricter in comparison with the regulations of mineral and spring water or the Israeli standard, but in other cases the values are lower. This can lead to variations in quality depending on source of supply: For example, while in mineral water a concentration of 0.01 mg/l cyanide and 45 mg/l is accepted, in drinking water supplied through pipes the accepted concentrations are 0.05 mg/l cyanide, respectively 70 mg/l.

Drinking water is a basic need of all human beings and should be supplied to the citizens equally with respect to quantity as well as quality. Therefore, it is hard to justify why drinking water should be supplied under various standards.

Can anybody really claim that the average citizen, having neither knowledge nor expertise in the field of standards for drinking water, can make a reasonable choice between the kinds of water he or she should consume?

5. Summary and Conclusions

According to the existing legal status, drinking water, which is not mineral water or spring water and supplied in bottles is not regulated under the People's Health Regulations (sanitary quality for drinking water). It is not regulated under the People's Health Regulations (mineral water and spring water), either, although both kinds of water are supplied in bottles, which makes them having the same marketing characteristics that may influence the quality of the water (e.g. the type of the bottle, amounts and types of the bacterial population). The Israeli standard setting standards for the quality of drinking water, which is not mineral or spring water supplied in bottles and containers, is not mandatory. Thus, the public does not have any certainty as to what the standards of drinking water in bottles should be. In cases where the main water supply will be restricted or unavailable due to security reasons, the most likely scenario will be the supplying of water in bottles or large containers. It is vital therefore that this method of supply be regulated to avoid large-scale diseases. We therefore recommend, to adopt the Israeli standard as mandatory for drinking water supply in bottles or alternatively to unify the standards of drinking water regardless of the method of supply.

The authors are from *The Levinson Environmental Law Firm*, which provides worldwide consulting and representation for corporations on environmental law. Additional information on environmental legal aspects can be found at www.environment.co.il.

SUBJECT INDEX

A
Aarhus Convention, 51, 54
Abiotic, 101, 102
Accounting, 148
Accumulators, 132
Acquis, 51
Action Plan, 11, 144
Activated Carbon, 134–137
Acute Trusts, 32
Adaptation measures, 78
Administration, 33, 37, 43–45, 48, 52, 72, 118, 145
Administrative Violations, 149
Adsorption, 69, 133, 135–137
Aesthetic effect, 127
Affected population, 79
Affordable water, 80
Ag^+, 129, 130
AgCl, 129
Age of water, 108
Agreements, 61, 62, 145, 149, 156
Agriculture, 4, 22, 23, 51, 53, 62, 63, 143, 147
  Agriculture by-product, 134, 135
  Agriculture development, 62
  Agricultural irrigation, 83
  Agricultural production, 50, 56
Agro-melioration, 88
Air temperature, 84, 116
Air-conditioners, 127
Airports Harbours, 32
AKOM, 48
Alarms, 17, 18, 94
  Alarm thresholds, 106, 112
Aldicarb, 4
Alibayramli, 142
Alibeyköy, 44, 45
Alloy, 127, 128
Almaz, 118
Alternative Sources, 20, 48, 144
  Alternative water sources, 20, 141, 143
Ammonia, 72, 134, 135, 137
Ammonium, 74, 98, 134, 135
Anaerobic digesters, 85
Anatolian, 43
Aniline, 134–137
Anions, 128, 129
Annual Total Water Demand, 44
Antenna grid, 117
Anthropogenic, 87, 88, 102
Anthropogenic catastrophes, 87
Apsheron Peninsula, 145
Aquatic life, 134
Aquifer, 22, 49, 63, 66, 78, 79, 82, 83, 85, 110
Arab, 61, 63
Aral Sea, 130, 131
Araz, 142
Arid regions, 59
ARIJ, 63
Armenia, 71, 73, 75, 88, 101, 134, 135, 137, 142, 146
Aromatic, 134, 137
Arsenic, 74, 106
Artesian well water, 130, 131
Artificial Recharge (AR), 110
  Artificial water catchments, 115
Anthropogenic load, 88
Anthropogenic impact, 88
Atmosphere, 116
Atmospheric fallout, 65, 68
Avalanches, 122
Average water consumption, 66
Awareness, 23, 52, 54, 55, 63, 153
Azerbaija, 141–156
"Azersu" Joint-Stock Company, 147, 151

B
Bacau, 53
Bacteria, 5, 96, 100, 128–130, 132, 161
Bacteria concentrations
  Bacteria killing time, 127
Bacterial cultures, 128
  Bacterial pollution, 127
  Bacterial regrowth, 129, 132
Baia Mare, 53
Baku, 142, 143, 151, 153
  Baku State University, 148, 152
Basin agreements, 149
Belgrade, 49
Benzene, 134, 137
Beta-contamination, 68
Biogas production, 85
Biogenes, 88
Biological
  Biological pathogens, 94
  Biological resources, 144

Biosphere Reserve, 49
Bioterrorism, 127, 132
Biotic, 101, 102
Biotoxines, 5
Black Forest
  Black Sea, 49–51, 54, 68
Blending water, 108
Bosphorus, 43, 44
Bottom sediments, 146
Boundary, 10, 43, 50, 51, 63, 107, 108
Braila, 53
Brasov, 53, 54
Bribin water, 82, 83, 85
Brundtland report, 80
BTEX, 137
Bucharest, 50, 52, 53, 55
Budapest, 49
Budiko classification, 142
Bulgaria, 52, 120, 121
  Bulgarian economy, 115
Buyukcekmece, 44–46
  Buyukcekmece Dam, 45
Bylaws, 144

C

^{137}Cs, 6, 66, 67–69
Calcareous, 85
Calcium, 6, 74, 134
Carbonate, 4
Carpathian Basin
  Carpathian Mountains, 49
Caspian Sea, 146
Catastrophes, 78, 87, 122
Catastrophic impacts, 78
Cations, 128, 129
Casualties, 3, 7, 88, 109
Caucasus 73, 142, 143
CCI4 solution, 135
Central Europe
  Central Java, 80, 81
  Central Sanitary, 129, 130
Centralised water infrastructure, 77
Centralized drinking water systems, 52
Centre of Epidemiology, 150
Cernavoda, 49
Chemical agents
  Chemical analysis, 102
  Chemical Procedures, 128
Chemical-Bacteriological Laboratory, 72
Chemicals, 3, 4, 7, 11, 13, 22, 89, 94, 96, 98, 106
Chernobyl, 65, 66, 68, 69
Chlorine, 5, 72, 95–97
  Chlorine residuals, 108
Cholera, 5, 79
Chromatography, 135–137
Circumpacific, 84
Cisterns, 82, 84, 85
Civil Protection, 31, 32
Cl^-, 74, 129
Clean dams, 69
  Clean water, 22, 43, 50
Clefts, 80, 81, 85
Climate, 79, 80, 119, 142
Climatic Changes, 24
Cluj, 53
Cobalt, 134, 162
Cole-Parmer, 128
Complex interactions, 79
Complexity, 4, 23, 25, 77
Comprehensive risk management, 79
Computer technology, 151
Conductivity, 95, 97, 98, 128
Conflict, 59, 61–63, 142
Construction projects, 148
Contaminants, 3–6, 11, 15, 19, 22, 53, 94–100, 133, 159
Contamination, 3, 6, 7, 11, 19–21, 24, 65–69, 97, 100, 105, 106, 113, 132
  Contamination Scenarios, 94
Contingency Planning, 25, 30, 106
Cooling water, 49, 127, 132
Copper, 74, 127–129, 134, 146, 161
Corrective action, 106, 113
Cosmos 1870, 118
Cost recovery, 149
Creeping threats, 78
Crisis, 24, 25, 33, 41, 42, 61, 101, 121
Critical assets, 100
  Critical Services, 13
Crop, 62
Cross-border
  Cross-boundary, 63
Cu^{2+}, 129
Cultivation, 128
Cyanic compounds, 4
Cyanide, 4, 106, 161, 163

D

Dam, 45, 61, 82, 83, 120, 121
Damage, 19, 20, 24, 33, 36, 48, 62, 78, 79, 85, 88, 100, 107, 143, 146
Danger, 59, 106
Danube, 49–51, 54, 55
Darlık Dam, 45
Databases, 106, 112
Decentralised water infrastructure, 86
Decision-making, 55, 145, 148

Decontamination, 6, 94
Deforestation, 78
Deliberate actions, xviii
Desalination, 51, 62, 127, 130, 131
Desertification, 145
Designer drugs, 4
Detection
 Detection Technologies, 94, 95, 98–100
Detector, 6, 94, 95, 137
Deterioration, 73, 88, 101
DHI, 111
Direct osmosis, 127
Directives, 51, 52, 55, 145, 154
Dirty bomb, 65
Diseases, 11, 79, 88, 159
Disaster, 3, 11, 22, 25, 48, 65–67, 72, 73, 77–79, 84, 87
Discharge, 49, 51, 73, 122, 146, 148–150, 154, 155
Disinfections, 128
Disruption, 13, 14, 32, 106, 107
Dissemination, 153
Distribution systems, 9, 17, 18, 20, 42–44, 48, 96, 97, 105, 106, 112, 146
Diversification, 77
Diversity, 49, 84
Djeiranbatan, 151
Dnieper river system
 Dnieper water system, 65, 68
Dobrogea, 50
Downstream utilities, 106
Drained, 50
Dráva, 49
Drinking water, 4, 11–14, 23, 38–40, 44, 49, 51, 52, 56
Drought, 3, 10, 21, 22, 48, 52
Drugs, 4, 79
Dry sanitation, 80
Duzdere Dam, 45

E

E. coli, 129–131
Early Warning, 11, 25, 105, 106
Earthquakes, 48, 78, 79, 84, 87
Eastern Mediterranean, 22
EBRD, 54, 153
Echo Technologies
 EchoSensor, 100
Ecolab, 153
Ecological balance
 Ecological blocks, 102
 Ecological catastrophes, 122
 Ecological factors, 102, 103
 Ecological monitory, 101
Economic policy, 49
Economical water use, 143, 155
Economy, 9, 77, 83, 88, 89, 102, 143–150, 152, 155
Ecosystem, 10, 11, 21–24, 59, 67, 75, 88, 101, 102, 104, 144, 146
Education, 63, 84, 151, 152, 159
Effective concentration, 128, 129
Efficient water use, 152
Effluent, 66, 79, 85
Egypt, 59
El Nino, 85
Electricity, 32, 38, 40, 54
Electrodes, 127, 128
Electrolytic process, 127
Elmalı, 44, 45, 46
Emergency
 Emergency Planning, 30–32
 Emergency Situations, 3, 11, 23, 24, 48, 67, 71–73, 132, 142–145, 155, 156
EN Technology Inc. (Korea), 130, 131
Energy, 10, 23, 30, 40, 50, 77, 107, 111, 127, 130–132, 143, 147, 152, 156
Enforcement, 13, 51, 71, 147, 156
Environment, 21, 32, 33, 37, 38, 53, 55, 68, 88, 102, 145–147, 150, 159
Environmental impact, 52, 75
 Environmental legislation, 51
 Environmental Protection, 24, 51, 54, 67, 107, 143, 149
Epidemics, 71, 84
ERS-1 satellite, 117, 118
ERS-2, 118
EU, 33, 49–52, 54, 55, 63, 151, 154
EU Directives, 51, 52, 151, 154
Europe, 31, 50, 54, 65, 66, 80, 110
European standard, 144, 156
Eutrophication, 73, 88, 146
Evaporation, 142
Exhaustion, 87
Extreme situations, 122

F

Failures, 3, 23, 48, 108, 109, 152
False alarms, 94
 False positive or negative responses, 105
$Fe(NO_3)_2 \cdot 6H_2O$, 128
 $Fe(NO_3)_3 \cdot 9H_2O$, 128
 Fe^{2+}, 128, 129
 Fe^{3+}, 128, 129
Fecal coliforms, 128
Fertilizers, 85
Fibroid sorbents, 127
Filters, 88, 133

Financial support, 143, 150, 151
Financing, 54, 79, 145, 152, 153
Fires, 78
Flash Floods, 22
Flood, 10, 21–23, 35–40, 42, 48, 49, 53, 54, 69, 78, 79, 116, 122, 143–145
Food Supply, 10
Forecasts, 102
Fresh Water, 5, 59, 110, 146
Fundulea City, 53
Fungi, 5

G
Gas-liquid, 136
Geographical information systems, 105
Geologic structure, 102
Georgia, 142, 146
Germany, 77, 127
Gold, 127–130
Good-quality water, 53
Great Britain, 150
Ground waters, 51, 53

H
Hazard categories, 78
Hazardous pollutants, 52
Health
  Health care system, 79
  Health hazard, 63, 106
  Health Regulations, 159–163
  Health Service, 18, 29, 32
Heavy metal, 134, 146
Helsinki, 61, 154
Herbicides, 4
Human health
  Human intervention, 78, 79
  Human rights, 33, 61
Human security, 61
Hungary, 49, 52
Hydro system
  Hydro-chemical statistical data, 156
  Hydro ecosystems, 101
Hydrogen bonds, 135, 136
Hydrographical network, 50
Hydrology, 23, 115, 116, 123
Hydrobiological
  Hydrological regime, 88
Hydrochemical, 67, 88, 156
Hygiene, 50, 67, 71, 88, 150

I
^{131}I, 6, 68
Ice-melted water, 115
ICTAF, 63
IDB, 153
Idjevan deposit, 136
Incubation time, 3
Indonesia, 41, 77, 80, 82–84
Indicators, 71, 72, 88, 95, 97
Industry, 10, 22, 31, 49, 53, 63, 105, 107, 141, 143, 151, 154, 155, 160
Inexhaustibility, 88
Information Center, 18
Infrastructure, 11, 15, 22–24, 36, 37, 52, 54, 55, 62, 77–80, 84–86, 93, 143
Inorganic, 53, 133
Insecticides, 4
Integrated water quality, 154
Internal irradiation, 67
International law, 60, 61
Intifada, 59, 62
Ion-exchange, 133, 135
Ions, 127–129, 134
IR spectroscopy, 137
Iran, 73, 142
Iraq, 61
Irrigation, 9, 12, 22, 62, 83, 88, 89, 144, 148, 149, 153
Israel, 9, 59, 61–63, 159, 160, 161
Istanbul, 43–48

J
Java, 77, 80, 81, 84, 85
Jordan, 59, 61, 62, 80
  Jordan river, 61
Justice, 61, 63, 80

K
Karst, 77, 80–83
karstic aquifers, 78, 79, 85
Kazandere Dam, 45
Kiev, 66, 68
Korea, 130, 131, 132
Kura river, 141–143, 146
Kuzuludere Dam, 45

L
Lake Sevan, 71, 73–75
  Lake Terkos, 43
Lead, 23, 29, 61, 74, 78, 79, 109, 110, 134, 161
Legal acts, 155
  Legal framework, 71
Legionella, 127–129
Legislation, 32, 33, 34, 51, 54, 55, 66, 67, 143, 144, 147–149, 155, 163
Liquid chromatography, 135–137
  Liquid effluents, 66
  Liquid waste, 149
London, 29, 30, 31, 32, 34

M
3-Methyltentanyl, 4
Magnesium, 74, 134, 161
Malaysia, 129, 132
Management Plan, 11, 12, 142
Man-made hazards, 78, 79, 85
Manufactured water, xvii
Marine water, 153
Marmara, 43
Membrane technology, 80
Mesta river, 115
Metals, 127, 128, 134, 146
Metal ions, 127–129, 134
Microflora, 134
Microorganisms, 84, 96
Micro-wave, 115, 116
Middle East, 59–62
Mineral Water, 160–163
Mitigation, 18, 19, 20, 25, 65, 69
Modernization, 54, 89
Monitoring, 71, 72, 74, 75, 95–99, 101, 102, 105, 106, 118, 122, 148
Municipalities, 52, 69, 148–150
Mustard, 5

N
Na_2S, 128
Na_2SO_4, 128
Nablus, 63
NaCl, 128
NASA, 118, 119
National plans, 51, 144, 154, 155
  National standards, 146, 151, 155
Natural Disasters, 11, 20, 22, 87
  Natural Hazards, 21, 24, 78, 79, 84, 85, 122
  Natural risk, 35
  Natural waters, 146
  Natural-climatic, 102
Nerve agents, 4, 106
NH_4, 74, 134, 135
Nickel, 74, 134, 146, 161
Nitrate Monitoring, 97
Nitrite, 74, 95, 97, 161
Nuclear
  Nuclear Power Plant (ChNPP), 65, 68, 69
  Nuclear stations, 49

O
Oceans, 80
OCMS, 93, 95, 98, 100
Oils, 53, 133
Ömerli, 44–46
Operational procedures, 94, 95, 108, 156
Optical biosensors, 100
Organic compounds, 133, 134
  Organic impurities, 135, 137
organoleptic indicators, 71, 72
Oslo, 61
Oxide, 4

P
Palestinian Authority, 59, 61, 62
Papucdere Dam, 45
Parathion, 4
Pathogens, 5, 94, 98, 99, 129
Pchelina dam, 115
Peres center, 62
Persistent contamination, 3
Phosphate, 95, 97
Physical security, 93, 100
Plant genetics, 62
Policy, 51, 61, 141, 144, 145, 147, 152–154, 156
Pollution, 21–23, 51–56, 66, 73, 78, 85, 87–89, 97, 109, 110, 127, 134, 143, 144, 146
Polyethleneimine, 136
Pond water, 129
Potable water, 13, 37, 38, 40, 66, 68, 94, 120, 160
Power plants, 49
Precipitation, 50, 81, 85, 115, 142, 162
Prevention measures, 79
Protection, 11, 13, 23, 24, 31, 32, 51, 53, 54, 66–69, 107, 143–150
Protocols, 17, 18
Public Health, 11, 13, 55, 72, 106, 112, 130, 159, 160
  Public security, 18, 77
  Public services, 52
Purification, 23, 69, 87, 88, 89, 135

Q
Quality Control (QC), 71, 72, 93, 95, 100

R
Radiation, 67, 68, 71, 134
Radioactivity, 17, 65, 68, 70
Radionuclides, 6, 66, 68, 69, 134
Rational use, 89, 90, 144, 146, 147, 149
Rapid Detection, 19, 93–100
Real time detection, 93, 95, 99, 100
  Real time response, 105, 106
Recovery Plan, 11
Regional cooperation, 61
Regulations, 31–34, 54, 67, 72, 85, 87, 127, 147–149, 162, 163

Rehabilitation, 53–55, 102, 141, 151, 153, 155
Reliability, 94, 95, 105, 107
Remote sensing, 115–123
Renewable, 59
Reservoirs, 4–7, 23, 37, 40, 43, 51, 65, 66, 68, 69, 72, 106, 108, 109, 134, 143
Resilience, 30, 77–79
Revival, 88
Rickettsial agents, 5
River, 35, 55, 115, 141, 146
Risk Analysis, 25
 Risk Assessment, 15, 25
 Risk Management, 24–25, 78, 79
Romanian waters, 51
Russia, 127, 130–32, 156

S

^{90}Sr, 6, 66–69
Sabotage, 13, 21, 24, 79, 80, 106
Saline, xvii
Salinisation, 22
Salmonella, 5, 128, 129
Salts, 53
Sandia Water Initiative, 98
Sanitary protection, 146, 149
Sanitation, 9, 11, 51, 80
SAR system, 117
Sazlıdere Damm, 45
Scenarios, 4, 25, 37, 38, 94, 95, 104, 106, 107, 109, 112
 Scenario Development, 24
Seawater, 51
Security, 4, 7, 10, 11, 13, 15, 17, 18, 22, 33, 40, 59, 61, 77, 79, 100, 159, 163
Sedimentation, 49
Self-purify, 88
 Self-rehabilitation, 87
Sensors, 98, 100, 106, 112
September 11, 3, 159
Sewage, 3, 22, 23, 53, 62, 63, 77, 80, 85, 88, 89, 146, 149, 150, 152, 154
Silver, 127–129
SO_4^{2-}, 128, 129,
Social assets, 79
 Social Goals, 11, 77, 93, 94, 105, 111, 145
 Social hazards, 78–80
Societal security, 77
Society, 36, 69, 77, 102, 145
Software, 17, 18, 105, 107
Soil samples, 153
Solar batteries, 130, 132
Sorbents, 133–135
Sorbing, 69
Source water, 105
Specord UV-spectrometer, 137
Spectral ranges, 116
Spores, 100
Spring Water, 160, 162, 163
Standards, 17, 18, 52, 55, 56, 62
State Water Protection, 148
Steel pipes, 66
Storage, 13, 53, 72, 85, 94, 106–108, 127, 132, 134
 Storage Tanks, 14, 94
Strategic Management, 21
Strategy, 51, 62, 65, 106, 147, 154, 156
Streams, 22, 44, 73, 80, 102, 119
Strouma river, 115
Studen Kladenets dam, 115
Sultanbahce Dam, 45
Surface waters, 51, 81, 101, 102, 143, 146, 147, 155
Surrogates Parameters, 95
Surveillance measures, 93, 100
Susceptibility, 77, 78, 86
Sustainable Development, 23, 25, 141, 143, 144
 Sustainable water management, 63, 77, 79, 86
Sweden, 151
Synthetic Aperture Radar (SAR), 116
Syria, 61
Systems analysis, 77, 83

T

Tap water, 72, 146
Technologies, 3, 7, 19, 22, 62, 69, 79, 86, 93–95, 98
Technology assessment, 83
 Technology transfer, 62, 142
Temperature, 52, 84, 88, 116, 119, 128, 137
Terkos Lake, 45
Thermophication, 88
Terror, 15, 19, 20, 65, 80, 94–97, 100, 159
Terrorism, 79
Tertiary, 133
Israeli standards, 160, 162
Threats, 11, 13, 15, 17–22, 78, 98, 105, 106
 Threats Categories, 17
Tigris, 80
Tools, 25, 79, 106
Topography, 43
Tornado, 78
Tourism, 22, 145
Toxicity, 4, 17, 18, 95
Toxins, 5, 6, 94, 100, 106

Training, 15, 54, 62, 104, 151, 152, 154
Transboundary, 10, 52, 55, 61, 63, 142, 144, 145, 154
Transport, 6, 9, 12, 29, 38, 51, 66, 69, 77, 79, 106, 108, 149
Transportation, 66, 77, 149, 150
Treated water, 66, 132
Trophia, 88
Tsunami, 41, 78, 87, 122, 127, 132
Turbidity, 39, 71, 96, 133, 141, 143, 152
Turkey, 9, 43, 73, 142
Typhoid – paratyphoid, 128–129
Typhus, 5, 79

U
Ukraine, 65–68
UN, 61
Underground water, 38, 39, 116, 130
UNDP/GEF, 153
UNESCO, 148, 152
UNICEF, 82
University of Karlsruhe, 81–83
Urban Ecosystem, 23, 24
Urbanization, 22, 36, 55, 88
USA, 65, 118, 130–132
Utilities, 11, 15, 20, 24, 29–32, 94, 96, 106, 109
UV spectrum, 136, 137
Uzbekistan, 127, 130

V
Valuing Water, 10,
Vibrio cholera, 128, 129
Viruses, 5
Volcanism, 78
Vulnerability, 11–15, 24, 25, 47, 77–79, 100, 159
VX, 5

W
Warning, 3, 7, 11, 25, 53, 54, 100, 105, 106, 109
Waste disposal, 68, 149
Wastewater, 23, 31, 44, 48, 51, 53, 56, 72, 79, 86, 88
Water authorities, 86
Water basins, 88–89, 150
Water Cleaning, 20
Water Code, 66, 144
Water disinfections devices, 128
Water distribution, 17, 18, 20, 44, 45, 47, 48, 54, 66, 72, 96, 106, 112
Water economy, 143, 148–152
Water economy facilities, 143
Water ecosystems, 67, 88, 144
Water Industry, 31, 155
Water infrastructure, 11, 54–55, 62, 77, 80, 84–86, 94
Water Law, 51
Water legislation, 55, 144, 155
Water management, 9, 11, 25, 55, 56, 61, 63, 77, 86, 142, 143, 146
Water network, 98, 100
Water policy, 51, 144, 145, 154
Water Protection, 68, 69, 146, 148, 153, 155
Water quality, 7, 10, 17, 18, 38, 40, 51, 63, 66, 69
Water reservoirs, 4, 7, 43, 143, 150, 152
Water right, 61
Water Safety, 89
Water Security, 11, 17, 18, 159
Water sources, 4, 5, 20, 62, 63, 73, 75, 83, 141, 143
Water Standards, 151
Water State Cadastre, 155
Water stock, 41, 51
Water storage, 94, 127, 132
Water supply, 3, 7, 9, 11, 12, 19–25
Water treatment, 17, 23, 37, 51, 53, 54, 62
Water users, 89
World Bank, 143, 144, 151, 153
World Health Organization, 150

Y
Yarmouk, 61
Yerevan, 72
Yugoslavia, 52

Z
Zeolites, 133–137
Zinc, 134

AUTHOR INDEX

Aleksanyan, Aida, 133
Burn, Sarah, 29
Coccossis, Harry, 21
Demirel, Burak, 43
Dimitriou, K., 21
Dror, Gil, 159
Einav, Amnon, 9
Gabrielyan, Bardukh, 71
Galatchi, Liviu-Daniel, 49
Hakobyan, Diana, 87
Hakopian, Christina, 87
Horsner, Marsha, 105
Ingeduld, Petr, 105
Isakov, Aleksander, 133
Jorgensen, P., Lingdaard, 105
Kallis, G., 21
Khaydarov, Rashid, 127
Khaydarov, Renat, 127
Kopfmuller, Jurgen, 77
Lehn, Helmut, 77
Levinson, Tzvi, 159
Mardirossian, Garo, 115
Negreanu, Jean, 93
Nguyen, Bruno, 35
Olsen, Roger, 127
Sack, Jacobo, 17
Schildt, Anne, 77
Sharan, Yair, 3
Tal, Abraham, 59
Theochari, Madeleine, 21
Torosyan, Gagik, 133
Tserunyan, Vardan, 71
Vardanian, Trahel, 101
Vaturi, Asher, 59
Verdiyev, Rafig, 141
Yenigun, Orhan, 43
Yuldashev, Bekhzod, 127
Zakaryan, Meri, 133
Zeman, Evzen, 105
Zlobenko, Borys, 65

Printed in the United States
81693LV00002BA/26